Site Investigation using Resistivity Imaging

To our parents

Site Investigation using Resistivity Imaging

Sahadat Hossain
The University of Texas at Arlington, Arlington, TX, USA

Golam Kibria
Arias Geoprofessionals, Inc., San Antonio, TX, USA

Sadik Khan
Jackson State University, Jackson, MS, USA

CRC Press
Taylor & Francis Group
Boca Raton London New York

CRC Press is an imprint of the
Taylor & Francis Group, an **informa** business

A BALKEMA BOOK

Published by:
CRCPress/Balkema
P.O. Box 447, 2300 AK Leiden, The Netherlands
e-mail: Pub.NL@taylorandfrancis.com
www.crcpress.com – www.taylorandfrancis.com

First issued in paperback 2020

ISBN 13: 978-0-367-57124-5 (pbk)
ISBN 13: 978-1-138-48593-8 (hbk)

Visit the Taylor & Francis Web site at
http://www.taylorandfrancis.com

and the CRC Press Web site at
http://www.crcpress.com

Typeset by Apex CoVantage, LLC

Library of Congress Cataloging-in-Publication Data
Names: Hossain, Sahadat, author. | Kibria, Golam, author. | Khan, Sadik, author.
Title: Site investigation using resistivity imaging / Sahadat Hossain, The
   University of Texas at Arlington, Arlington, TX, USA, Golam Kibria, Arias
   Geoprofessionals, Inc., Arlington, TX, USA, Sadik Khan, Jackson State
   University, Jackson, MS, USA.
Description: Leiden : CRC Press/Balkema, [2018] | "CRC Press/Balkema is
   an imprint of the Taylor & Francis Group, an informa business." | Includes
   bibliographical references and index.
Identifiers: LCCN 2018015379 (print) | LCCN 2018016096 (ebook) | ISBN
   9781351047609 (ebook) | ISBN 9781138485938 (hardcover : alk. paper)
Subjects: LCSH: Geotechnical engineering—Technique. | Earth resistance
   (Geophysics)—Measurement. | Soils—Analysis—Technique. | Electrical
   impedance tomography. | Engineering geology.
Classification: LCC TA705 (ebook) | LCC TA705 .H67 2018 (print) | DDC
   620.1/91042—dc23
LC record available at https://lccn.loc.gov/2018015379

Contents

Preface

Subsurface investigation is the most important phase of any civil engineering construction or development activity. The geologic conditions can be extremely complex, variable, and subject to change with time. Soil test borings are traditionally used for subsurface exploration; however, this method provides information on a point at different depths and does not yield a general overview of the site. Therefore, considering the uncertainty and variability of the subsurface in between soil test borings or test locations, geotechnical engineers often become overly conservative to avoid unsatisfactory outcomes.

Inadequate site investigations can lead to unexpected/unwanted surprises and circumstances during construction, which may cause delays and extra costs, or may even lead to failure of a project. Since limited soil test borings cannot provide comprehensive site information, one can argue that more soil test borings can be conducted to get additional site information. However, in many cases, that becomes too expensive for the project and still may not provide comprehensive information in between test points. Reliable and cost-effective designs are vital to construction planning and are only possible with complete and representative site investigations.

Subsurface investigation is also required for routine maintenance and evaluation of geohazard potentials of geo-infrastructures (i.e., earth slopes, levees, earth dams, etc.). In many cases, due to limited accessibility of drill rigs on a sloping surface, soil test boring is limited in earth slopes, levees, and earth dams.

Subsurface investigation is also an essential component of geoenvironmental engineering projects. The determination of solid waste characteristics and moisture contents during the operation and performance monitoring of a bioreactor landfill operation is an example of a geoenvironmental application to a subsurface investigation. The moisture content of solid waste in a bioreactor landfill is typically determined by using bucket auger drilling, which is not cost effective and provides information at the testing point only, followed by laboratory testing.

The use of geophysical methods for site investigations offers the opportunity to overcome the limitations and inherent problems of conventional site investigation methods. The use of a geophysical method, electrical resistivity imaging (often called as resistivity imaging, RI), is gaining notable recognition from global engineering and construction communities for geotechnical/geoenvironmental site investigations. Advantages of RI over conventional site investigation methods include: (1) a continuous image of subsurface conditions, (2) coverage of a large area within a short time, (3) low cost, (4) observations of site heterogeneity and zones of high moisture contents, (5) quick and easy data processing, and (6) is not operator dependent. Because of these benefits, the use of RI has increased significantly in recent years.

RI is usually performed by a geologist or geophysicist whose primary focus is obtaining qualitative information. Their explanations and viewpoints are different in certain cases from what geotechnical and geoenvironmental engineers need for design or analyses of a site condition. Since there is a growing trend for geophysical testing to be integrated into geotechnical site investigations, along with other conventional *in-situ* tests, it is essential to close the gap that currently exists between geotechnical engineering and geophysics in data interpretation. If the gap is not properly bridged, it can be difficult for geotechnical engineers to understand and interpret results presented by geophysicist.

At present, there is no book available to engineers that correlates resistivity imaging results with geotechnical/geoenvironmental engineering parameters. ***This book aims to bridge the gap that currently exists between the geotechnical/geoenvironmental and geophysical engineering community.*** The contents of this book are presented in a way that geotechnical and geoenvironmental engineers will be able to interpret the geophysical data and utilize the information for their design. It is a comprehensive handbook for the application of RI in geotechnical and geoenvironmental site investigations.

The authors would like to take the opportunity to thank everyone who helped them complete this book. Special thanks to:

- Previous graduate students who worked with Dr. Hossain and were directly or indirectly involved during the resistivity imaging in the field: Dr. Jubair Hossain, Dr. Shahed Rezwan Manzur, Dr. Asif Ahmed, Dr. Jobair Bin Alam, Dr. MD Ishtiaque Hossain, Josh Hubbard, Kenta Fujimto, and Dr. MD Zahangir Alam. We are extremely thankful to them for their important service.
- Dr. Huda Shihada, Dr. Hossain's former graduate student currently working as Environmental Engineer at AECOM, for allowing us to include part of her dissertation in Chapter 5 of the book. We greatly appreciate Dr. Shihada's unconditional help.
- Current researchers and graduate students who helped us with different aspects of this book: Dr. Asif Ahmed (Research Assistant Professor), Anuja Sapkota, Prabesh Bhandari, Tanvir Imtiaz, and Fouzia Hossain. We are very thankful to them for their hard work, dedication, and service.
- Vance Kemler, previous general manager, and David Dugger, facility manager, Solid Waste & Recycling Services at the City of Denton, Texas, for their continuous support during the resistivity imaging for their landfill operation, and for their encouragement to complete this book.
- Texas Department of Transportation (TxDOT) for continuous support during the geohazard investigations of various slopes, using resistivity imaging.
- The researchers who allowed to include some of their work in the book. We greatly appreciate it.
- Ms. Ginny Bowers, our official reviewer/editor, for checking grammar and other aspects of the book.
- Dr. Janjaap Blom, editor at the Taylor and Francis Group, for his keen interest in the book.

Sahadat Hossain, Ph.D., P.E.
Golam Kibria, Ph.D., P.E.
Sadik Khan, Ph.D., P.E.

About the authors

Dr. Sahadat Hossain received his bachelor's degree in Civil & Environmental Engineering from Indian Institute of Technology (IIT), Bombay, India in 1994 and Master's Degree in Geotechnical Engineering from Asian Institute of Technology (AIT), Bangkok, Thailand, in 1997. He received his PhD degree in Geo-Environmental Engineering from North Carolina State University (NCSU), USA in 2002. He worked as a Civil/Geotechnical engineer in Bangladesh, Thailand, Singapore and USA before joining the University of Texas at Arlington (UTA) in 2004 as a faculty. Dr. Hossain is a professor at the Department of Civil Engineering at UTA and founding director of Solid Waste Institute of Sustainability (SWIS), an Organized Research Center of Excellence at UTA. Dr. Hossain has more than 20 years' experience in Geotechnical /Geoenvironmental Site Investigations, Design, Construction and Operation. His site investigation experience includes soil test boring, pressure meter test, dilatometer test, and resistivity imaging. Dr. Hossain's design experiences includes shallow and deep foundation design for building and bridge, excavation support system and retaining structures, cut and cover tunneling, slope stability analysis, design and construction of drilled shaft, contiguous bored pile wall, secant pile wall and diaphragm wall. Dr. Hossain's experience also includes design, operation and management of sustainable waste management system, including bioreactor landfill design, landfill gas to energy projects, landfill mining, reuse of recycled materials for civil engineering infrastructure projects. He authored and co-authored one book, published more than 100 of papers in prestigious journals and conference proceedings.

Dr. Golam Kibria is a Geotechnical Engineer at Arias Geoprofessionals, Inc. in San Antonio, Texas. He received his Master's and Ph.D. degree from the University of Texas at Arlington. He has work experience in Bangladesh and United States. His research and work experience include: subsurface investigation using conventional and geophysical methods (i.e. electrical resistivity imaging, seismic survey), sustainable slope stabilization, deep and shallow foundation, laterally loaded pile, retaining wall, and water infrastructures. He published various articles in ASCE and Elsevier Journals. He is an author of "Sustainable Slope Stabilization using Recycled Plastic Pin" book published by Taylor and Francis. Dr. Kibria is a professional engineer in the State of Texas.

Dr. Sadik Khan received his bachelor's degree in Civil Engineering from Bangladesh University of Engineering and Technology, Dhaka, Bangladesh in June 2007. He received his Master's in Civil Engineering on summer 2011 and Ph.D. in Civil Engineering on fall 2013. Dr. Khan is working as an Assistant professor at the Jackson State University from fall 2015. He worked more than 10 years in different projects with on the site investigation and slope stability analysis, Forensic Investigation of MSE wall failure, performance evaluation of highway slopes and MSE wall, Unknown Bridge Foundation and Geophysical methods such as Resistivity Imaging. He worked on the sustainable slope stabilization technique using recycled plastic pin, implemented in several highway slopes, evaluated performance and developed the design protocol. Dr. Khan is a member of ASCE GI Unsaturated Soil, ASCE GI Shallow Foundation and TRB AFP 30-Soil and Rock Properties Committee. He coauthored more than 30 peer reviewed publications in the prestigious Journals and Conference Proceedings. Dr. Khan is a licensed engineer in the State of Texas and State of Mississippi.

Chapter 1

Introduction

1.1 General

Subsurface investigation is the most important phase of any civil engineering construction or development activity. Since the geologic conditions can be extremely complex, variable, and subject to change with time, soil test borings and *in-situ* tests are employed to obtain subsoil information. Traditional *in-situ* tests provide information of a point at different depths, but do not yield a general overview of the site. Therefore, considering the uncertainty and variability of subsurfaces in between the soil test borings or test locations, geotechnical engineers often become overly conservative to avoid unsatisfactory outcomes.

Inadequate site investigations often lead to unexpected/unwanted surprises during construction, which may cause delays and extra costs, or may even lead to the failure of a project. Since limited soil test borings in a few points cannot provide comprehensive site information, one can argue that more soil test borings can be conducted to get additional site information. However, in many cases that becomes too expensive for the project and still may not provide comprehensive information in between test points. Reliable and cost-effective designs are vital to construction planning and are only possible with complete and representative site investigations.

Subsurface investigation is also required for routine maintenance and evaluation of geo-hazard potentials of geo-infrastructures (i.e., earth slopes, levees, earth dams, etc.). In many cases, due to limited accessibility of drill rigs on a sloping surface, soil test boring is limited in earth slopes, levees, and earth dams.

Subsurface investigation is also required for many geoenvironmental engineering projects, such as determining waste sample characteristics and moisture content during the operation and performance monitoring of a bioreactor landfill. The moisture content of solid waste in a bioreactor landfill is determined by using bucket auger drilling, which is not cost effective and provides information at the testing point only (Manzur *et al.*, 2016), followed by laboratory testing. Therefore, a comprehensive site investigation method, which can provide a continuous subsurface profile and geotechnical and geoenvironmental parameters, is required.

1.2 Current subsurface investigation methods

At present, various *in-situ* tests and drilling soil test borings are employed to determine subsoil properties. The common *in-situ* tests are the standard penetration test (SPT), cone penetration test (CPT), pressure meter (PMT), flat dilatometer (DMT), and vane shear tests (VST).

Furthermore, various standard *in-situ* test methods are developed by local public agencies to better and more comprehensively capture the geologic condition of that area. Some examples are Texas cone penetration (TCP) test, modified penetration test, Iowa borehole shear test, etc. The objective of each type of test is to measure the subsoil response and correlate it to the material geotechnical properties such as strength and stiffness. A brief description of some widely used *in-situ* tests are presented in the following subsections.

1.2.1 Standard Penetration Test (SPT)

Standard penetration test (SPT) is the most common *in-situ* test method. It is conducted during drilling to determine the soil resistance at various depths. Additionally, disturbed samples are obtained using a split spoon during SPT tests. The standard test procedures for SPT are described in ASTM D1586.

SPT is conducted in conjunction with soil test drilling. A 63.5 kg hammer is pounded in a repeated manner from a height of 0.76 m to achieve a total of 450 mm in three successive increments (150 mm each). The SPT blow count requires penetrating the first 150 mm, which is known as "seating" blow counts. The number of blows required to penetrate successive 300 mm are known as SPT blow counts, i.e., N-value. When the penetration of 150 mm is not achieved, typical industry practice is to record penetration for 50 blow counts. It should be noted that the hammer may encounter bedrock, very dense gravel, or an obstacle such as a boulder. If this occurs, it is recommended that the boring be extended below this depth, under the direction of a geotechnical engineer. Typically, SPT tests are performed at 0.76 m intervals at depths up to 3 m, then 1.5 m intervals thereafter. However, depending on the subsoil conditions and type of project, SPT is sometimes conducted continuously up to 4.5 m or deeper. When wet rotary/wash boring is used under the ground water, the head of water in the borehole must be maintained to avoid borehole instability.

SPT is widely used in various subsurface conditions and weak rocks. It is easily available, fast, and can be performed at any depth. Since SPT is widely used, many studies have been performed to develop correlations between soil parameters and SPT.

SPT is highly dependent on the type of equipment and competence of the operator. According to ASTM D4633, calibration of energy efficiency for the rig and a skillful operator are required. Instrumented strain gages and accelerometer measurements can be used to standardize the hammer energy. The average energy is about 60% in the United States. While using SPT, the observed blow counts may need to be standardized, depending on the hammer type, hammer energy, etc.

1.2.2 Cone Penetration Testing (CPT)

Cone penetration test (CPT) is considered one of the most popular and effective *in-situ* tests. It provides continuous profiling of subsoils at a particular test location. It is a fast, economical, and productive *in-situ* test method. ASTM D3441 and ASTM D5778 provide details of CPT test methods for mechanical and electric systems, respectively. In this test, a cylindrical steel probe is pushed into the subsoil at a constant rate of 20 mm/sec., and the resistance to penetration is recorded in terms of sleeve and tip resistance. A standard CPT probe consists of tip with an apex angle of 60 degrees and a 35.7 mm diameter. The area of a standard friction sleeve is 150 cm^2. As per the ASTM standard, the test can be performed using a 43.7 mm diameter and 200 cm^2 friction sleeve. CPT is a fast and economic method for continuous profiling of subsoils, it is applicable to very soft clays to dense sands, and it is not operator dependent. However, this method is not suitable in gravel, boulder, or rock subsurfaces, and samples cannot be obtained by using this method in subsoils that are composed of them. With the advancement of technology, various sensors such as piezocones, resistivity cones, acoustic cones, seismic cones, vibrocones, cone pressuremeters, and lateral stress cones can now be added to the probe to determine project specific parameters.

1.2.3 Pressuremeter Test (PMT)

The pressuremeter test is an *in-situ* load test conducted within a borehole. A long cylindrical probe is expanded radially at a specific depth where the test is performed. The diameter and length-to-diameter ratio of standard probes range from 35 to 73 mm and 4 to 6, respectively. Water or gas is usually used as a fluid medium in soils, and hydraulic oil is used in rock materials. A stress-strain curve is developed by interpreting the amount of fluid required to expand the probe within the borehole. Standard test procedures and calibrations for pressuremeter tests are described in ASTM D4719.

The pressuremeter was introduced by the French geotechnical engineer and entrepreneur, Louis Menard, in 1955. The arrangement, calibration, and testing were complex in the early pressuremeters. Simpler commercial pressuremeters (Texam, Oyo, and Pencel) are now available to perform tests more conveniently. Simple commercial test setups include a monocell probe with a displacement-type screw pump for inflation. Currently, there are four different pressuremeters available: Menard-type, self-boring, push-in, and full displacement. The pressuremeter provides four independent parameters, i.e., lift-off stress, and corresponds to the total horizontal stress, elastic zone, plastic zone, and limit pressure (related to bearing capacity). Although a pressuremeter test is technically sound and provides *in-situ* soil/rock parameters, it is a time consuming, expensive, and delicate test method. Because it is a very complicated test procedure, it requires a high level of expertise in the field.

1.2.4 Dilatometer Test (DMT)

The flat dilatometer test (DMT) provides at-rest lateral stresses, elastic modulus, and shear strength of *in-situ* soils, using the pressure readings from an inserted plate. The DMT was first introduced by Silvano Marchetti in Italy and is currently used in more than 40 countries.

The DMT *in-situ* test equipment consists of a tapered stainless-steel blade with 18° wedge tip. The wedge tip is pushed into the ground at intervals of 200 or 300 mm, at a rate of 20 mm/s. The approximate dimensions of the blade are 240×95×15 mm. Wire tubing through drill rods or cone rods is used to connect the blades to a readout pressure gauge. Two pressure readings, A-reading and B-reading, are determined by inflating a 60-mm diameter flexible steel membrane that is located on one side of the blade. An A-reading is considered as the contact pressure where the membrane becomes flush with the blade face (zero displacement); B-reading is considered as the expansion pressure that corresponds to 1.1 mm outward deflection at the center of the membrane. Movement is detected by using a tiny spring-loaded pin at the membrane center that transmits to a buzzer/galvanometer at the readout gauge. Due to low moisture content, nitrogen is preferred for the test; however, carbon dioxide or air can also be used.

The standard test procedure for DMT is described in ASTM D6635. Two calibrations are recommended before the actual testing to determine the correction factors for the membrane in the absence of soil. Pressure reading "A" is taken about 15 seconds after insertion, and

"B "is obtained 15 to 30 seconds later. The membrane is quickly deflated, and the blade is pushed to the next test depth after completion of pressure reading "B". The DMT is a simple, robust, quick, and economical *in-situ* test method. This test is not operator dependent; however, it requires calibration for local geologies and relies on correlations. Additionally, it is often difficult to perform DMT in dense sands and hard clays.

1.2.5 Vane Shear Test (VST)

The vane shear test (VST) is conducted to estimate the peak and remolded undrained shear strength of soft-to-medium stiff clays. A four-bladed vane is pushed into the clayey soil and rotated slowly at a rate of 0.1 degrees per second. While rotating, the resisting torque evolving from soil shearing is measured. As noted, two shear strengths are determined from VST: peak shear strength and remolded shear strength. The peak torque is related to the peak shear in a cylindrical failure surface by a constant and depends on the dimensions and shape of the vane. After obtaining the peak torque, the vane is rotated about ten (10) times to determine the torque associated with remolded shear strength. The sensitivity of the clayey soil is determined by calculating the ratio of peak strength to remolded shear strength. The standard test procedure is described in ASTM D2573.

Sensitivity of the soil is useful in measuring the drivability analysis of pile. It is an effective method for estimating undrained shear strength less than 50 kPa for clays. This is an economical and rapid test method; however, it is only useful in soft-to-medium stiff clays and is only applicable when just undrained shear strength is required.

1.3 Limitations of the conventional methods

The available *in-situ* tests provide information across depths at a specific location. A general overview or continuous image of subsoils cannot be obtained using these methods. Additional limitations are:

1   Although the direct method of sampling is capable of determining the different properties of the soil, it is expensive to collect the large number of soil samples that are necessary for accurate determination of the soil profile and heterogeneity.
2   In many cases, the time-dependent moisture variation information is crucial; for example, in landslide studies. To monitor the moisture variation, several instrumentation techniques that provide useful information have been developed. However, this data can only be obtained for several points, and it is expensive to collect time-dependent information, using conventional soil borings.
3   The existing technics are very good for collecting information along the 1D plane; however, it is difficult to get entire view of the subsurface through the borehole data. As a result, geotechnical engineers sometimes perform linear interpolation to collect the missing data, which can be misleading.

1.4 Electrical resistivity imaging method for site investigations

The use of geophysical methods for site investigations offers the opportunity to overcome the limitations and inherent problems of conventional site investigation methods. The use of a geophysical method – electrical resistivity imaging (often called as resistivity imaging) – is gaining notable recognition from the global engineering and construction communities for site investigations. Resistivity imaging (RI) is a non-destructive, fast, and cost-effective method of site investigation and soil characterization. Advantages of RI over conventional methods include: (1) a continuous image of subsurface conditions, (2) coverage of a large area within a short time, (3) low cost, (4) observations of site heterogeneity and zones of high moisture content, (5) quick and easy data processing, and (6) not being operator dependent. Because of these benefits, the use of RI has increased significantly in recent years. It is one the most convenient, available techniques for preliminary subsurface investigations in geotechnical and geoenvironmental applications. A typical RI result is presented in Figure 1.1. This RI was performed as a part of the subsurface investigation for retaining walls to be designed and constructed for the Texas Department of Transportation (TxDOT).

It is evident from the result that RI provides continues images of the subsurface. The effectiveness of the RI method in subsurface investigations can be further explained by considering three different boring locations with the RI results presented above. Let's consider two borings within the array length of the RI results, which is typical for any retaining wall project. Three different cases of boring locations are presented in the RI images in Figure 1.2.

According to conventional industry practices, linear or curvilinear interpolations are used to determine the subsurface profile. The possible consequences of area-wide interpolation of subsoil profiles in these three cases are presented in the following table.

Another important aspect of RI in engineering applications is the possibility of obtaining engineering parameters and their implications in analysis and design. RI is performed primarily by a geologist or geophysicist whose background and first degree is in either physics or geology. Geophysics may be their second degree. Their primary objective in performing the test is to obtain qualitative information. They have very little knowledge of the civil engineering contractual constraints within which civil and construction engineers must work while evaluating operational efficiency and geohazard potentials, and their explanations and viewpoints are, in some cases, different from what geotechnical and geoenvironmental engineers need for their design or analysis of a site condition. There is a growing trend for geophysical testing to be integrated into site investigations, along with other conventional *in-situ* tests, in a way that is difficult for geophysicists to understand.

According to current practices, only qualitative subsurface information can be obtained from RI. Since engineers utilize subsurface investigation results in their designs and analyses, they are interested in determining geotechnical and geoenvironmental parameters from RI. Quantification of geotechnical and geoenvironmental properties has become an important issue for rigorous use of RI in engineering applications. The correlations of different geotechnical and geoenvironmental properties with RI allows geotechnical and geoenvironmental engineers to interpret the geophysical data and utilize the information for their analyses and designs.

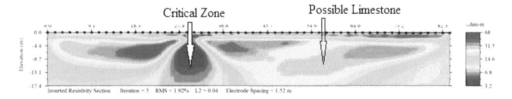

Figure 1.1 Resistivity image showing the possible limestone and critical zones in subsurface (Hossain *et al.*, 2014), with permission from ASCE

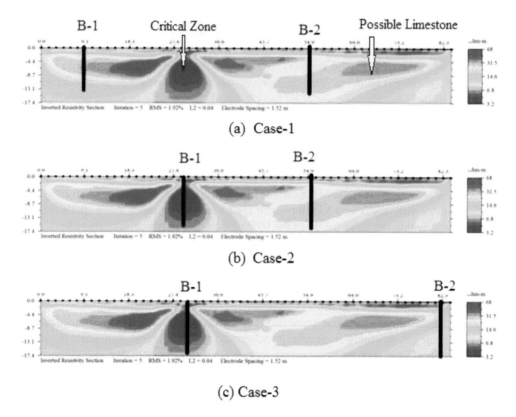

(a) Case-1

(b) Case-2

(c) Case-3

Figure 1.2 Different boring locations with resistivity imaging (RI) results

Table 1.1 Possible consequences of area-wide interpolation of subsoil profiles

Cases	Possible consequence	Influence in the design and construction
Case–1	Missing the critical low resistivity zone	Inadequate design. Possible failure or foundation distress
Case–2	Misinterpretation of the extent of limestone	Conservative design. Increases the project cost
Case–3	Missing the limestone	Overly conservative design. Increases the project cost

The RI method provides subsurface information based on the electrical resistivity of soils. The electrical resistivity of soil depends on several geotechnical properties such as moisture content, unit weight, porosity, pore fluid conductivity, clay fraction, surface charge, fabric structure, tortuosity, temperature, etc. (McCarter, 1984; Kalinski and Kelly, 1993; Abu Hassanein *et al.*, 1996; Fukue *et al.*, 1999; Yang, 2002; Rinaldi and Cuestas, 2002; Giao *et al.*, 2003; Ekwue and Bartholomew, 2010; Kibria and Hossain, 2012; Kibria and Hossain, 2017). A general model to explain the variation of electrical conduction with soil properties is not available due to the inherent complexity of the soil-water matrix and interconnectivity among the influential parameters. Researchers have presented a number of empirical models and theoretical equations based on laboratory experiments and simplified assumptions. The

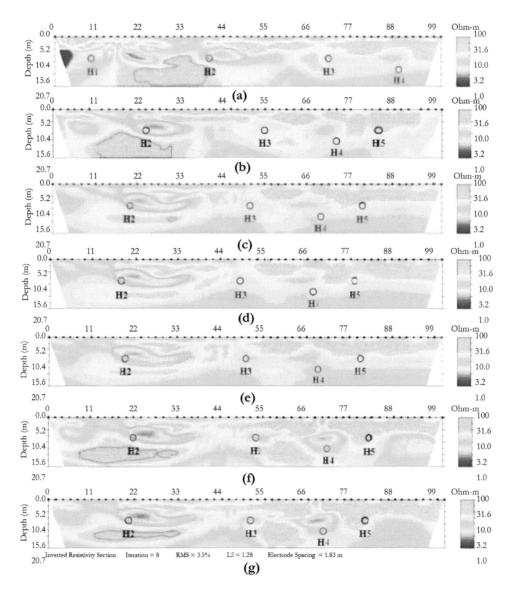

Figure 1.3 (a) Baseline on 05/22/2009, (b) one day after recirculation of 18920 liters (5006 gallons) on 7/29/2009, (c) one day after recirculation of 19470 liters (5151 gallons) on 8/26/2009, (d) one day after recirculation of 18940 liters (5010 gallons) on 9/26/2009, (e) one week after recirculation of 18940 liters (5010 gallons) on 10/2/2009, (f) one week after recirculation of 37870 liters (10019 gallons) on 10/23/2009 and (g) one week after recirculation on 22715 liters (6009 gallons) on 11/4/2009 (Manzur *et al.*, 2016[1])

1 Reprinted from Waste Management,, 55, Manzur, S. R., Hossain, M. S., Kemler, V., & Khan, M. S. , Monitoring extent of moisture variation due to leachate recirculation in an ELR/bioreactor landfill using Resistivity Imaging, 8-48, Copyright (2016), with permission from Elsevier.

available mixing models describe resistivity as a function of pore fluid conductivity and surface conductance. The experimental methods to determine these parameters are often time and cost intensive. Moreover, characterization tests on pore water and surface charge are not typically performed in a conventional geotechnical investigation. Therefore, practical, applicable models are required to understand the effects of geotechnical parameters on resistivity and will be helpful in quantifying geotechnical properties from resistivity measurement.

In addition to geotechnical site investigations, RI is being considered as an effective technique in geoenvironmental applications. A study conducted by Manzur *et al.* (2016) indicated that the RI method is an effective technique for evaluating the moisture variations and performance of a leachate recirculation system of a bioreactor landfill. The study was conducted at the City of Denton Landfill in Denton, Texas. A horizontal recirculation pipe was selected, and the vertical and horizontal extent of the moisture variations during the recirculation process was monitored for 2.5 years, using the RI technique. The moisture accumulation process across a horizontal pipe is presented in Figure 1.3.

Electrical resistivity depends on the moisture content, composition, unit weight, porosity, pore fluid composition, temperature, decomposition, and organic content of municipal solid waste (Shihada *et al.*, 2013). However, the changes in resistivity are significant with the variations of moisture content, temperature, and porosity, and the effects of the other parameters on resistivity have not been well established for solid waste (Grellier *et al.*, 2007). Since the application of RI has increased significantly in the municipal solid waste area, it is important to document the parameters affecting the electrical resistivity of solid waste.

At present, there is no book available to engineers that summarizes pertinent details of RI in site investigations for geotechnical and geoenvironmental applications. Therefore, this book aims to bridge the gap that currently exists between the geotechnical/geoenvironmental and geophysical engineering communities, enabling geotechnical and geoenvironmental engineers to interpret the geophysical data and utilize the information for their analyses and designs. This book presents electrical resistivity measurement techniques in the laboratory and field, effects of geotechnical and geoenvironmental parameters on electrical resistivity of soils and municipal solid waste, development of practical and applicable models and their application to the field scale, and case studies. The chapters in this book are organized as follows:

Chapter 1 presents a general description of electrical resistivity imaging in the application of geotechnical and geoenvironmental investigations.

Chapter 2 describes background of the electrical resistivity principle in subsurface investigation and measurement methods in the field and laboratory.

Chapter 3 presents the geotechnical parameters that affect the electrical resistivity of geomaterials.

Chapter 4 presents available mixing models for soil electrical resistivity, their applicability and limitations, the development of practical and applicable models for soils, and correlations of resistivity with the geotechnical properties of clayey soils.

Chapter 5 presents the municipal solid waste properties that affect the electrical resistivity of soils and a model that correlates electrical resistivity with municipal solid waste properties.

Chapter 6 presents some case studies that illustrate the applications of RI in geotechnical and geoenvironmental studies.

Chapter 2

Background

Electrical resistivity of geomaterials and measurement methods

Electrical resistivity was employed by Gray and Wheeler in 1720 in the field of geology to determine the conductivity of rocks (Jakosky, 1950; Van Nostrand and Cook, 1966). In 1746, Watson ascertained that a subsurface has the ability to conduct electricity (Van Nostrand and Cook, 1966). Robert W. Fox conducted a study on sulfide ore to determine the existence of conductivity and, by using a copper electrode, determined the presence of a natural current within the sample (Van Nostrand and Cook, 1966). The application of the DC current to quantify resistivity was performed by Conrad Schlumberger in 1912, and it was reported as one of the most successful experimental approaches to electrical resistivity (Aizebeokhai, 2010). In the United States, the idea was developed by Frank Wenner in 1915 (Aizebeokhai, 2010), and since then, the method has undergone significant improvements. To comprehend the heterogeneity and to provide accurate images of subsurfaces, different electrode combinations and inversion models are currently being utilized. With the advancement of modern techniques, it is now possible to obtain images of a subsurface within a very short time.

2.1 General principle of electrical conductivity and resistivity

The fundamental physics utilized in the measurement of electrical resistivity of any material is Ohm's law, where voltage is the product of the current and resistance of the material. A schematic of current flow through a cylindrical section is presented in Figure 2.1.

The current density (J) is a microscopic vector quantity and can be defined as the electric current per unit of a cross-sectional area. The current is represented by the dot product of the current and cross-sectional area as follows:

$$I = \int J \cdot dA \tag{2.1}$$

If the electric field vector is E, then the potential difference ΔU can be written as

$$\Delta U = -\int E \cdot dl \tag{2.2}$$

dl is the element along the integration path of electric field vector E.

In a uniform electric field, the above two equations can be substituted in Ohm's law:

$$U = IR(\text{Ohm's law}) \tag{2.3}$$

$$E.l = J \cdot A \cdot \frac{\rho \cdot l}{A} \tag{2.4}$$

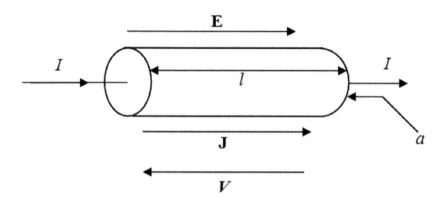

Figure 2.1 Flow of current through a cylindrical cross section

$$E = J.\rho, Or\ E = \frac{J}{\sigma} \tag{2.5}$$

where ρ is the resistivity of the material, which is a function of resistance, length of conduction path, and cross-sectional area of the conductive material. σ is the conductivity of the material that is reciprocal of resistivity.

The DC resistivity can also be regarded as the retardation of low-frequency alternative current (AC) signals. Therefore, magnetic properties of the material can be ignored at low-frequency conditions, and Maxwell's equation can be written as:

$$\nabla.E = \frac{1}{\varepsilon_0}q \tag{2.6}$$

$$\nabla \times E = 0 \tag{2.7}$$

where E is the electric field in the vector form, ε_0 is the dielectric permittivity of the free space $(8.854\times10^{-12}$ F/m), and q is the charge density. The electric field (E) can be presented as the gradient of electric potential (U) as follows:

$$E = -\nabla U \tag{2.8}$$

In a three dimensional space (x, y, z)

$$\nabla.E = \frac{1}{\varepsilon_0}q(x,y,z) \tag{2.9}$$

$$E = -\nabla U(x,y,z) \tag{2.10}$$

Substituting the vector **E**, the expression can be presented as

$$\nabla^2 U(x,y,z) = -\frac{1}{\mu_0}q(x,y,z) \tag{2.11}$$

The Dirac delta function can be employed to describe the continuity equation of a point in 3D space.

$$\nabla.J\left(x,y,z,t\right)=-\frac{\partial q\left(x,y,z,t\right)}{\partial t}\partial\left(x\right)\partial\left(y\right)\partial\left(z\right)\tag{2.12}$$

Based on the vector form of Ohm's law, the above equation can be rearranged to

$$-\nabla.[\sigma(x,y,z)\nabla U(x,y,z)]=\frac{\partial q(x,y,z,t)}{\partial t}\partial(x-x_s)\partial(y-y_s)\partial(z-z_s)\tag{2.13}$$

where x_s, y_s and z_s are the coordinates of the injected current source. The source of the current can be represented by considering an elemental volume ΔV:

$$\frac{\partial q\left(x,y,z,t\right)}{\partial t}\partial\left(x-x_s\right)\partial\left(y-y_s\right)\partial\left(z-z_s\right)=\frac{I}{\Delta V}\partial\left(x-x_s\right)\partial\left(y-y_s\right)\partial\left(z-z_s\right)\tag{2.14}$$

The current I can be injected through a point source (i.e., electrodes in the field condition). In an isotropic non-uniform 3D medium, a partial differential equation of electric potential can be developed by replacing the equation 2.13 with 2.14.

$$-\nabla.\left[\sigma(x,y,z)\nabla U\left(x,y,z\right)\right]=\frac{I}{\Delta V}\partial\left(x-x_s\right)\partial\left(y-y_s\right)\partial\left(z-z_s\right)\tag{2.15}$$

This is the basic equation for the determination of potential distribution in the subsurface under the application of current from a point source.

2.2 Electrical conduction in geomaterials

Electrical conduction in a particulate medium generally occurs by the movement of ions through electrolytic pore water in the void and surface charge (Bryson, 2005). In coarse-grained soil, conduction is largely electrolytic and depends on the interconnected pore space, granular skeleton, electrolyte conductivity, and degree of saturation (Santamarina et al., 2001). However, surface charge is an important parameter in the electrical conduction of clayey soils. Clay particles possess charge deficits due to substitution of ions in crystal structures and acid-based reactions of silanol-aluminol (Si-O-H and Al-O-H) groups with water. Adjacent cations are attracted to the clay particles to counterbalance the net negative charge. The density of cations is high around the solid surface; however, concentrated cations try to diffuse to equalize concentration throughout the structure. The diffusion phenomena are restricted by the negative electrical field of clay particles, and anions are moved away because of the negative force of the particles. As a result, relatively mobile ions, consisting of both positive and negative charges, exist contiguous to the adsorbed layer. The combined charged surface and distributed charge surface is known as the electrical double layer. The plane along which the counter ions are strongly adsorbed with negative charge of particles is designated as the Stern layer. An application of external electrical fields results in the charges separating in the diffuse double layer, along the Z-potential plane (Revil et al., 1998; Rinaldi and Cuestas, 2002). Therefore, electrical conduction in clayey soil depends on bulk fluid and surface conductivity. A schematic of the locations of the diffuse double layer (DDL), the Stern layer, and precipitated ions in clays is presented in Figure 2.2.

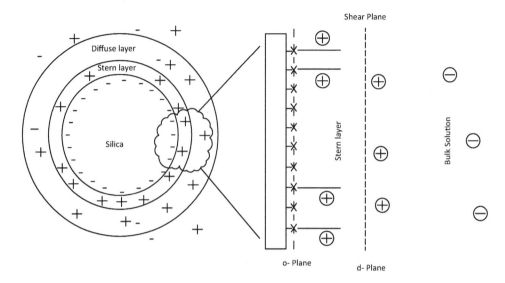

Figure 2.2 Location of diffuse double layer (DDL), Stern layer, and precipitated ions in clays

The interaction of water with clayey soil also plays a pivotal role in electrical conduction. In dry conditions, adsorbed cations are tightly held together by the negative charge of clay particles. After neutralizing the net negative charges of clay particles, excess cations exist as salt precipitates. Precipitated salts go into the soil-water solution in the presence of moisture. Previous studies indicated that the electrical and thermodynamic properties of adsorbed water are different from free water (Holtz and Kovacs, 1981). Revil *et al.* (1996) also emphasized the role of chemical reactions in clay surfaces in the presence of water. According to the study, a particle surface with silanol group can be dissociated to positive or negative charges (SiOH $\leftrightarrow$ SiO$^-$+H$^+$/ SiOH+ H$^+\leftrightarrow$ SiOH$_2^+$), based on the chemical reaction in the presence of water.

2.3 Measurement of electrical resistivity

Electrical resistivity of soils can be measured in the field or in the collected soil samples in the laboratory. In the field, electrodes are placed in the ground and connected to a resistivity meter. In the laboratory, soil samples are compacted in a box, and resistivity is measured using a resistivity meter. Detailed descriptions of laboratory and field measurement procedures are presented in the following sections.

2.3.1 Laboratory scale

The electrical properties can be studied in the laboratory by using direct current (DC) or alternative current (AC). During material characterization AC, frequencies ranging from low Hz to microwave can be used. The frequency factor is not considered due to the application of direct current in the DC method. The working principle in the DC method is associated with

Ohm's law, where voltage drops across the electrodes are measured by applying an electric current. Resistivity tests can be performed with two and four electrode configurations. A brief description of both methods is presented in the following sections.

Two-electrode system

Two-electrode measurements of electrical resistivity are described in the ASTM G187 standard test method. A two-electrode soil box, current source, resistance measuring equipment, and electrical connections are utilized in the tests, as presented in Figure 2.3. In this method, the same electrodes are used for the current application and voltage measurements. The two-electrode soil box should be made of insulated and durable material to avoid short circuiting during experiments. Two end plates, constructed with polished and corrosion-resistant metal, can be used for current flow and voltage measurements. It is recommended that the resistivity at 15.5°C (288.65 K) be measured according to the ASTM G187 standard and the following equation:

$$\rho_{15.5} = \frac{(24.5+T)}{40}\rho_T \tag{2.16}$$

here, $\rho_{15.5}$ = resistivity corrected at 15.5°C (288.65), ρ_T = measured resistivity at medium temperature, T = temperature during experiment.

Santamarina *et al.* (2001) indicated possible errors in two-electrode resistivity measurements. The electrical conduction in the electrode and cable generally involves electron flow; however, the flow of current is mostly ionic in the soil. Therefore, charge accumulation may occur in the soil-metal interface. This phenomenon is called polarization and is the main source of error in a two-electrode measurement. In addition to polarization, the presence of an air gap at the interface or the presence of a non-uniform electric field and possibility of chemical reaction may cause additional errors in this method.

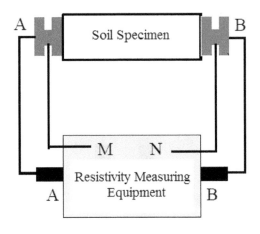

A, B = Current supply
M, N = Voltage measurement

Figure 2.3 Two-electrode electrical resistivity measurement system

Four-electrode system

A current is applied, using two electrodes that are located at both ends of the soil resistivity box, and the potential drop is measured between the two points. This method has an advantage over the two-electrode measurement system because the potential is determined within the sample, which is away from the charge transfer process of the current electrodes, meaning that polarization can be largely avoided. In addition, this method measures the voltage within the sample; thus, the actual electric field of the sample can be encountered during the tests. As the voltage and current electrodes are different, the possible effect of a chemical reaction on the electrical resistivity measurement may not be significant in this case. The experimental setup of four-electrode measurements is presented in Figure 2.4.

2.3.2 Field scale

Electrical resistivity measurement has been utilized in the investigation of near-surface geology since the early 20th century, but it has only become popular in recent years, with the improvement of test methods and data processing. At present, geoelectrical measurements are a useful tool in geophysics, soil science, hydro-geological studies, and environmental and geotechnical engineering (Aizebeokhai, 2010; Hossain *et al.*, 2010).

A current (I) is injected through the current electrode C1 in the isotropic homogenous half-space of earth (Figure 2.5). The electric potential decreases inversely with the increase of distance from the current source. The current distribution follows an outward radial direction through the shell area of $2\pi r^2$, perpendicular to equipotential lines. The potential for one electrode can be defined as:

$$\phi = \frac{\rho I}{2\pi r} \tag{2.17}$$

Typically, two current electrodes are used as positive and negative ends in a conventional resistivity survey. In this case, the potential distribution has a symmetric pattern around the

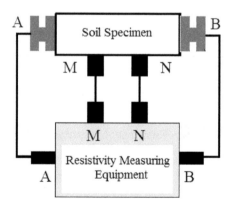

A, B = Current supply
M, N = Voltage measurement

Figure 2.4 Four-electrode electrical resistivity measurement system

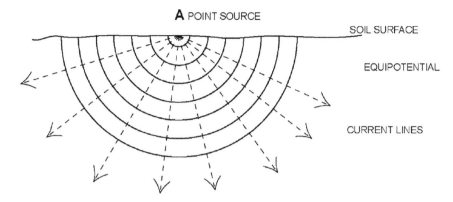

Figure 2.5 Current flow and equipotential lines due to a point source (Samouëlian *et al.*, 2005)[1]

vertical plane, centered at the midpoint of the electrodes. The potential for two-electrode configuration can be defined as:

$$\phi = \frac{\rho I}{2\pi}\left(\frac{1}{r_1} - \frac{1}{r_2}\right) \tag{2.18}$$

r_{c1} and r_{c2} are the distances of the measured point from the first and second current electrodes. The equation can be extended for a four-electrode system as:

$$\phi = \frac{\rho I}{2\pi}\left(\frac{1}{r_1} - \frac{1}{r_2} - \frac{1}{r_3} + \frac{1}{r_4}\right) \tag{2.19}$$

A schematic of a four-electrode system is presented in Figure 2.6.

Although the equations presented here are applicable for homogenous isotropic medium, actual field surveys are associated with the anisotropic and inhomogeneous subsurfaces. Therefore, an apparent resistivity is calculated from the measured current and potential, according to the following equation:

$$\rho_a = \frac{k\Delta\phi}{I} \tag{2.20}$$

where, k is a geometric factor that depends on the arrangement of the four electrodes and can be expressed as:

$$k = \frac{2\pi}{\left(\dfrac{1}{r_{c1p1}} - \dfrac{1}{r_{c2p1}} - \dfrac{1}{r_{c1p2}} + \dfrac{1}{r_{c2p2}}\right)} \tag{2.21}$$

The apparent resistivity can be defined as the electrical resistivity of a homogenous subsurface medium that will provide the same resistance in the electrode configuration. It can be

1 Reprinted from Soil and Tillage Research, vol 83, issue 2, A. Samouëlian,I. Cousin,A. Tabbagh,A. Bruand,G. Richard, Electrical resistivity survey in soil science: a review, 173-193., Copyright (2005), with permission from Elsevier.

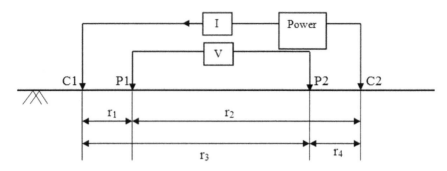

Figure 2.6 Four-electrode measurement of resistivity in the field

considered as a weighted average of the resistivity of the subsurface volume under the four electrodes. The apparent resistivity depends on the electrode arrays, and inversion modeling is required to determine the true resistivity from apparent measurements (Aizebeokhai, 2010).

One-, two-, and three-dimensional resistivity surveys can be performed in the field. The choice of survey depends on the site conditions and project objectives. A brief description of one-, two-, and three-dimensional resistivity surveys are presented as follows:

One-Dimensional Resistivity Survey: One-dimensional resistivity measurement at the field condition is generally known as vertical electrical sounding (VES). The spacing is successively increased to obtain resistivity at the deeper section in this method. This method provides resistivity variations along the vertical direction and does not consider the changes in horizontal resistivity (Loke, 1999).

Two-Dimensional Resistivity Survey: Two-dimensional, multi-electrode arrays can provide continuous 2D images of the subsurface. The current and potential electrodes are placed at a fixed spacing, and measurements are progressively moved from one end to another, as illustrated in Figure 2.7. Based on the measurement of apparent resistivity, a 2D pseudo-section can be developed. Thereafter, inversion modeling is performed on the measured apparent resistivity to obtain continuous 2D images of the subsurface (Samouelian *et al.*, 2005).

Three-Dimensional Resistivity Survey: Three dimensional resistivity surveys can provide robust information about the subsurface. There are two methods for determining the 3D resistivity profile of subsurface: (1) quasi-3D and (2) actual 3D resistivity survey. In a quasi-3D resistivity survey, different 2D parallel pseudo-sections are combined to evaluate the 3D profile of the investigated area. The measurements should be performed in X and Y directions to obtain an actual 3D resistivity profile (Arjwech, 2011).

2.4 Electrical resistivity inversion modeling

When a current is injected through a point source, the basic equation for the determination of potential distribution in the earth can be expressed as:

$$-\nabla.\left[\sigma(x,y,z)\nabla U(x,y,z)\right]=\frac{I}{\Delta V}\partial(x-x_s)\partial(y-y_s)\partial(z-z_s) \tag{2.22}$$

A forward modeling method can be employed to solve the equation. Typically, the finite difference and finite element modeling are utilized to solve the equation for 2D and 3D resistivity measurements; the analytical method can be used for 1D resistivity surveys.

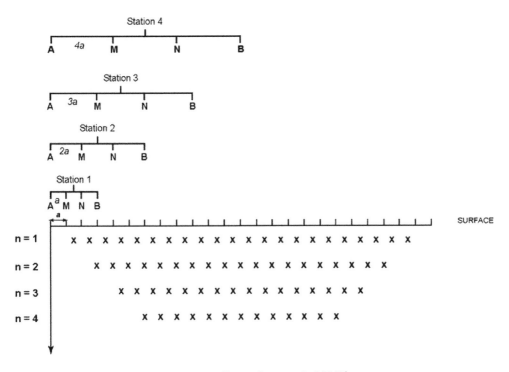

Figure 2.7 2D resistivity measurement (Samoulien *et al.*, 2005)[2]

During inversion of resistivity measurements, a model is determined that can provide a similar response to the actual values. The model consists of a set of parameters which are the physical quantities estimated from the observed data (Loke, 1999). In a subsurface resistivity distribution, forward modeling can be utilized to provide theoretical values of apparent resistivity. Typically, finite difference and finite element modeling are used to calculate the theoretical apparent resistivity. Eventually, inversion methods determine a subsurface model whose responses match with the measured quantities under certain conditions. Details of mathematical procedures for inversion modeling are presented in the study of Loke (1999) and review of Arjwech (2011). A typical flow diagram of inversion modeling is presented in Figure 2.8.

2.5 Electrical resistivity array methods

Typically, Wenner, dipole-dipole, Schlumberger, pole-pole, and pole-dipole arrays are utilized in the one-, two-, and three-dimensional resistivity surveys. Brief descriptions of the arrays are presented in the following subsections.

2 Reprinted from Soil and Tillage Research, vol 83, issue 2, A. Samouëlian,I. Cousin,A. Tabbagh,A. Bruand,G. Richard, Electrical resistivity survey in soil science: a review, 173-193., Copyright (2005), with permission from Elsevier.

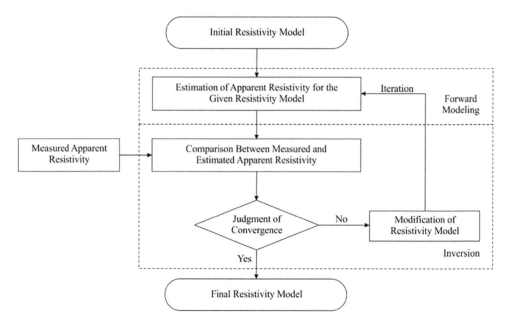

Figure 2.8 Algorithm of inversion modeling (Arjwech, 2011)

2.5.1 Wenner array

The use of the Wenner array has become popular because of the extensive research of the University of Birmingham. The array is more sensitive to vertical changes than to horizontal variations in resistivity. Typically, a Wenner array can evaluate horizontal structures; however, its performance is poor in mapping narrow vertical structures. Moreover, the Wenner array is preferred for surveys where substantial noise is anticipated in the field condition. The electrode configuration of a Wenner alpha array is presented in Figure 2.9.

2.5.2 Dipole-dipole array

The dipole-dipole array is characterized by low electromagnetic coupling; therefore, it is regarded as an efficient method in field surveying (Loke, 1999). The spacing between current and potential electrodes is the same in this array as illustrated in Figure 2.10. It is more sensitive to variations in the horizontal direction than to the vertical changes in resistivity. The dipole-dipole array is a popular method for imaging vertical structures.

2.5.3 Schlumberger array

The schematic of a Schlumberger array is presented in Figure 2.11. This array is more sensitive to vertical resistivity than to horizontal variations. The horizontal coverage can be decreased with an increase in the electrode spacing (Aizebiokhai, 2010). In recent years, a relatively new method, the Wenner–Schlumberger method, has been utilized for resistivity sounding.

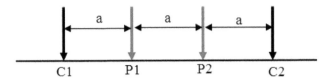

Figure 2.9 Wenner array

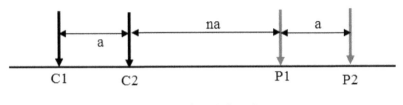

$$K= \pi n\ (n+1)\ (n+2)\ a$$

Figure 2.10 Dipole-dipole array

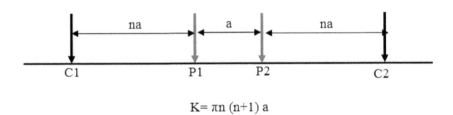

$$K= \pi n\ (n+1)\ a$$

Figure 2.11 Schlumberger array

2.5.4 Pole-pole array

The conventional pole-pole array consists of one current and one potential electrode (C1 and P1). The second current and potential electrodes are located at a distance of more than 20 times the spacing between C1 and P1. This method is not used as often for resistivity sounding as the Wenner array, Schlumberger array, or dipole-dipole array. It has the highest coverage area in horizontal and vertical directions; however, the resolution of the obtained image is not satisfactory because of the large spacing between the electrodes. The electrode configuration in the pole-pole array is presented in Figure 2.12.

2.5.5 Pole-dipole array

The pole-dipole method is characterized by high signal strength and less sensitivity in response to telluric current. This is an asymmetric method and has a relatively high horizontal coverage. Due to asymmetric electrode configuration, the pole-dipole method is divided into two groups: forward- and reverse-pole-dipole array. The schematic of the electrode configuration in a pole-dipole array is presented in Figure 2.13.

A summary of characteristics of different surveys is presented in Table 2.1.

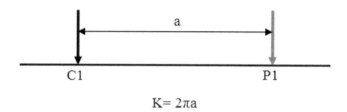

$$K = 2\pi a$$

Figure 2.12 Pole-pole array

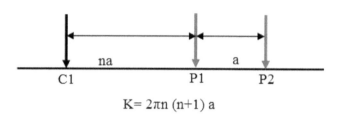

$$K = 2\pi n (n+1) a$$

Figure 2.13 Pole-dipole array

Table 2.1 Summary of characteristics of different arrays (Arjwech, 2011)

	Wenner	Wenner–Schlumberger	Dipole-Dipole	Pole-Pole	Pole-Dipole
Sensitive to horizontal structures	4	2	1	2	2
Sensitive to vertical structures	1	2	4	2	1
Depth of investigation	1	2	3	4	3
Horizontal data coverage	1	2	3	4	3
Signal strength	4	3	1	1	2

Poor sensitivity = 1 and high sensitivity = 4

2.6 Resistivity imaging method

Resistivity imaging (RI) is also known as electrical resistivity tomography (ERT). In a traditional resistivity survey, four electrodes are placed in the ground and a low-frequency or direct current is applied. The spacing of the electrodes depends on the scope of the test and the intended surveying methods. A current is applied across two electrodes, and voltage is measured across two other electrodes. RI is an advanced development of the traditional resistivity survey. In RI, enhanced data collection and inversion modeling provide continuous two-dimensional or three-dimensional resistivity models. The resolution of image and depth of investigation depend on the electrode spacing and test methods. Once field data collection is completed, data can be processed in computers to obtain resistivity images of the investigated area. During data processing, inversion parameters should be chosen carefully to obtain a representative image of the investigated area. Typically, equipment manufacturers provide a guideline for choosing inversion parameters. RI is considered as one of the most effective site investigation tools in clayey subsoil where GPR is not a viable option. Although resistivity has been used by geologists, soil scientists, and archeologists for many years, its application in geotechnical engineering is new.

2.7 Advancement in RI technics: single channel vs multi-channel system

Single-channel instruments require electrodes to be reconfigured after each measurement. This methodology is time consuming and laborious. Newer equipment has multiple channels that include multiple electrodes, where measurements are taken through each channel. The SuperSting R8 resistivity meter, produced by Advanced Geosciences, Incorporated (AGI), is equipped with eight channels. Therefore, for each current injection, the system utilizes nine electrodes to collect eight different potential difference measurements (AGI, 2006). The variations of data measurements of single channel and multichannel systems are presented in Figure 2.14.

Multiple-channel equipment has to receive instruction on the proper triggering sequence of electrodes. This information can be programmed by manual entry or uploaded from a coded command file. The sequencing information considers the array style and information pertaining to the electrode locations/electrode address during each measuring sequence.

There are no theoretical limits to the depth of penetration. However, for tomography applications, practitioners can assume that the depth of penetration is approximately 15% to 25%. The electrode spacing should not be greater than twice the size of the object or feature to be imaged. The design of the survey (i.e., survey run length, electrode spacing, and array type) directly impacts the depth of penetration and resolution (AGI, 2006).

2.8 Roll-along survey

For continuous profiling, the roll-along survey technique can be utilized to continuously profile the subsurface along a common survey line. Figure 2.15 presents both 2D and 3D roll-along patterns. As shown, a segment of electrodes is detached from the original survey line and relocated to the end of the cable system, effectively advancing the survey along the

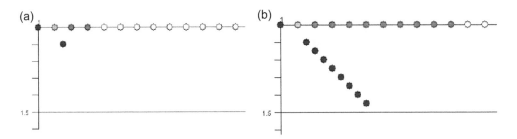

Figure 2.14 (a) Single-channel instrument; (b) eight-channel instrument (AGI, 2006)

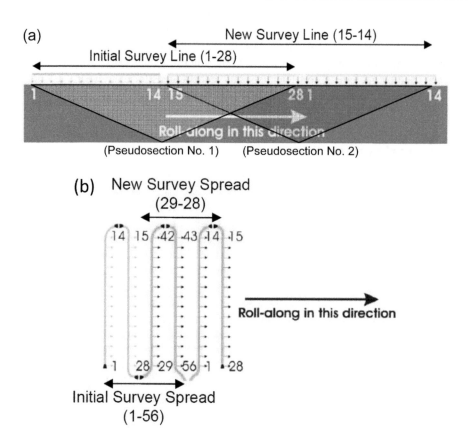

Figure 2.15 Example of a roll-along for (a) Two-dimensional survey; (b) Three-dimensional survey (AGI, 2006)

desired imaging path. As the entire line will not be advanced, not all of the readings will be new. In Figure 2.15 (a), the overlapping triangular patterns represent the respective field data points or pseudo-section generated during each survey. Based on the instrument and survey design, these readings may either be repeated or disregarded after acknowledging that the area has already been measured. Using multiple cables, the amount of data overlap is reduced and more of the survey can be performed with fewer movements. However, a decrease in data overlap will increase the void left below the two overlapping pseudo-sections, as presented in Figure 2.15 (AGI, 2006).

Chapter 3

Geotechnical properties affecting electrical resistivity

3.1 General

This chapter presents a description of the geotechnical parameters that influence the electrical resistivity of soils. A number of studies have been performed to determine the relationship between electrical resistivity and soils, and moisture content has been established as one of the major factors that causes changes in soil resistivity. However, the effect of unit weight degree of saturation, organic content, pore water composition, cation exchange capacity, specific surface area, and mineral and geologic formations are also significant. Detailed discussions on the factors influencing electrical resistivity are presented in the following subsections.

3.2 Geotechnical properties affecting electrical resistivity of soils

3.2.1 Moisture content

The engineering behavior of soils, specifically cohesive soils, is highly influenced by the presence of moisture. Therefore, the amount of moisture present in the soil is one of the basic parameters that a geotechnical engineer needs to know. Electrical conductivity occurs through the precipitated ions in the soil pore water. Free electrical charges cause a reduction in electrical resistivity under the application of the electric field. Therefore, electrical resistivity decreases with an increase in moisture. It is reported that the rate of reduction in resistivity with moisture is significantly below 15% moisture content (Samouelian *et al.*, 2005).

Pozdnyakov (2006) divided electrical resistivity vs. the natural logarithm of moisture content curve into various segments, as presented in Figure 3.1. The segments of the curve were designated as adsorbed water, film water, film capillary water, capillary water, and gravitational water. According to the author, electrical resistivity decreases rapidly with the increase of moisture content in the adsorption water zone. Although the ions of water molecules are immobile in the adsorbed water zone, the dipolar water creates a conductive path for electrical current. Thus, the electrical resistivity decreases substantially with the increase of moisture in the adsorption zone. However, the rate of reduction decreases in the film water zone because of the increase in Van der Waals's force. When the maximum possible thickness of water film develops, pore water goes from film to fissure. The molecular attraction force is higher than the capillary force in the film capillary water zone; therefore, electrical resistivity decreases less dramatically in the film capillary and capillary water zones. In the gravitational

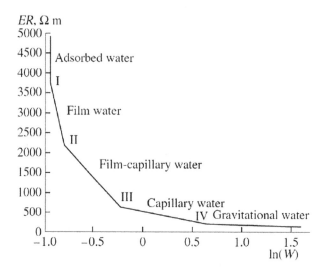

Figure 3.1 Soil moisture and electrical resistivity relationship (Pozdnyakov, 2006)[1]

water zone, mobility of electrical charges becomes independent of the movement of water molecule ions, and electrical resistivity is almost independent of water content.

Due to high surface activity, clay particles are easily hydrated by water and are surrounded by several layers of moisture films. The water adjacent to the crystal structure of the clayey soil is known as adsorbed water. The physico-chemical phenomena of clayey soils largely depends on this water. Mojid and Cho (2006) described the variations of electrical conductivity (EC) with moisture based on the formation stages of diffuse double layer (DDL). Tests were conducted on marine clays, bentonites, and bentonite-sand mixtures. The results indicated that the EC of the test specimens was low at their dry state and increased to a maximum value at high moisture contents. After that, the EC of the soil remained constant over a small range of moisture content; however, a decrease in EC occurred with further increase in moisture. According to the authors, the variations of EC with water content were associated with the developmental stages of DDL in the samples. At low water content, the cations in the clay surface are exposed to the moisture and adsorbed to develop DDL. However, the thickness of DDL is very thin at this moisture condition and clay particles are not in electrical contact through their DDLs. With the increase of moisture, the cations and adsorbed water surround the clay particles. At this condition, electrical conductivity increases because of the presence of a continuous pathway. However, the DDLs start dissociating from each other at very high moisture contents.

Ozcep *et al.* (2009) presented a study to determine the relationship of soil resistivity and water content in Istanbul and Golcuk, Turkey. Electrical resistivity was measured using vertical electrical sounding (VES) in 210 points of two sites. In addition, soil test borings were conducted for the collection of samples. The soil resistivity and moisture contents ranged from 1 to 50 Ohm and 20% to 60%, respectively. Two exponential equations correlating

1 Reprinted by permission from Springer Nature: Springer Nature Eurasian Soil Science, Relationship between water tension and electrical resistivity in soils, A. I. Pozdnyakov, L. A. Pozdnyakova, L. O. Karpachevskii, Copyright 2006.

moisture content with resistivity were developed for the Istanbul and Golcuk areas, as presented below:

$$W = 51.074e^{-0.0199R}, R^2 = 0.76\left(for\ Istanbul\right) \tag{3.1}$$

$$W = 47.579e^{-0.0158R}, R^2 = 0.75\left(for\ Golcuk\right) \tag{3.2}$$

Kibria and Hossain performed a study on the effects of moisture contents in compacted and undisturbed samples between 2010 and 2014. In 2012, soil resistivity tests were conducted at varied moisture contents, keeping the unit weight constant in fat clay samples. The designations of the test specimens are presented in Table 3.1.

Moisture contents varied from 10% to 50% during tests. Soil samples were compacted at their optimum dry unit weights. It was observed that soil resistivity decreased almost linearly up to moisture content around 20% for all soil samples. The average reduction in soil resistivity was 13.8 Ohm-m for the increase of moisture from 10% to 20%. Maximum variation was observed in soil sample 2. Resistivity decreased from 21.1 Ohm-m to 3.3 Ohm-m for the increase of moisture content from 10% to 20% in sample 2. Moreover, it was found that soil resistivity results were almost constant after 40% moisture content. The observed soil resistivity ranged from 2.1 to 2.42 Ohm-m at 50% moisture content in the soil samples. The variations of resistivity with moisture content of soil samples are presented in Figure 3.2.

Kibria and Hossain (2012) conducted soil resistivity tests in the dry state to identify the influence of surface charge of clay in the absence of moisture. It was observed that there was no flow of current through the soil in the dry state.

Another study conducted by Kibria in 2014 indicated that the soil resistivity decreased as much as 6.8, 4.8, and 3.5 times their initial values for the increase of moisture contents from 10% to 30% at 13.4 kN/m³ dry unit weight in Ca-bentonite, CL, and CH specimens, respectively, as presented in Figure 3.3. However, substantial variations in resistivity were not observed in the moisture range of 30% to 40% in these samples. The resistivity decreased from 501 to 46 Ohm-m for a 10% to 30% increase in moisture in the kaolinite sample. Moreover, resistivity reduced from 46 to 33 Ohm-m in the 30% to 40% moisture range in this specimen.

The variations of resistivity in response to gravimetric moisture contents for undisturbed soil samples obtained by Kibria (2014) are presented in Figure 3.4. Although initial moisture contents were not similar in the undisturbed soil specimens, reduction in resistivity followed similar trends. In undisturbed soil specimens, resistivity values ranged from 3.2 to 49.4 Ohm-m for the increase in moisture content from 7% to 31.3%.

Determination of a representative subsurface moisture profile is difficult because of the heterogeneity and complex hydro-geologic system of soil. Time domain reflectometry

Table 3.1 Classification of soil samples according to Unified Soil Classification System (USCS)

Designation	Soil Type	Liquid Limit	Plasticity Limit
SAMPLE 1	CH	73	28
SAMPLE 2	CH	61	27
SAMPLE 3	CH	79	28
SAMPLE 4	CH	58	26

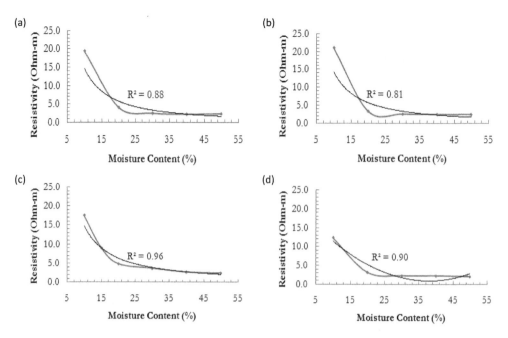

Figure 3.2 Variation of soil resistivity with gravimetric moisture content (a) Sample 1 (γ=14.7 kN/m³) (b) Sample 2 (γ=15.2 kN/m³) (c) Sample 3(γ=15 kN/m³) (d) Sample 4 (γ=14.9 kN/m³) (Kibria and Hossain, 2012), with permission from ASCE

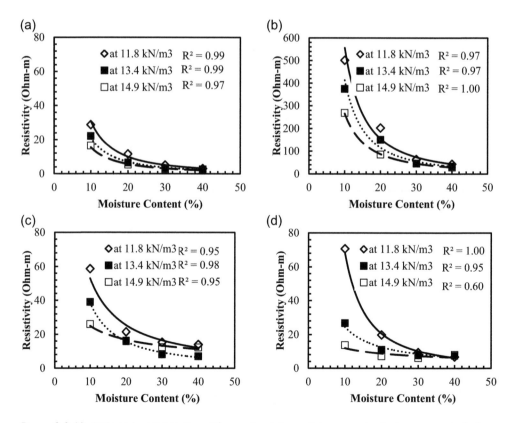

Figure 3.3 Variations in resistivity with gravimetric moisture contents in compacted clays at different dry unit weights (a) Ca-bentonite, (b) Kaolinite, (c) CL, and (d) CH

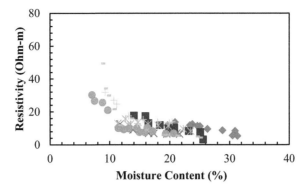

Figure 3.4 Variations in resistivity with gravimetric moisture contents in undisturbed samples

(TDR), neutron probes, gypsum blocks, tensiometers and gravimetric scaling are commonly used to measure moisture distribution in subsoil. However, some of these methods have certain operational limitations. For an example, determination of moisture using TDR is possible within a region of only tens of centimeters around the probe (Goyal *et al.*, 1996). In contrast, electrical resistivity can be utilized to determine the moisture condition of the subsurface. Several studies indicated that resistivity decreases with the increase of soil moisture. This phenomenon led to several researchers' efforts to quantify the moisture content of soil from resistivity in the laboratory and field.

Crony *et al.* (1951) described a methodology to determine soil moisture, using the electrical resistance method. The measurement was based on three relationships: (1) the suction of the water in the absorbent and moisture content of the absorbent, (2) moisture content of the absorbent and the resistance of the gauge, and (3) the suction of water in the soil and moisture content of the soil. Plaster of Paris and high alumina cement were used as absorbent materials. It was observed that the electrical resistance gauges could be used to determine the soil suction and soil moisture; however, their reliability as a soil moisture meter was doubtful because of the disturbance of the soil. According to the study, it is important to calibrate electrical gauges to obtain precise results. The measurements of suction and moisture content of the absorbent were identified as major problems in this method because very small differences in mixing and curing of absorbents influences the results significantly.

Schwartz *et al.* (2008) conducted a study to quantify field-scale moisture content by using the 2D electrical resistivity imaging (ERI) method at the Virginia Tech Kentland experimental farm, Montgomery County, Virginia. ERI and time domain reflectometry (TDR) were used simultaneously to obtain resistivity and moisture content. The 1D resistivity profile was determined from 2D ERI, using EarthImager software. The coefficients of Archie's law were numerically optimized for the quantification of moisture content from 1D resistivity. The proposed model utilized extractable cations to represent the role of pore water conductivity in developing Archie's law. The use of extractable cations eliminated difficulties in measuring extracted pore water resistivity. It was observed that the model provided useful results for determining meter-scale moisture heterogeneities compared to small-scale variations.

Brunet *et al.* (2010) conducted research to obtain water deficits from electrical resistivity tomography (ERT) in Southern Cevennes, France. From February 2006 to December 2007, more than ten ERTs were performed on the study area, and volumetric water contents were

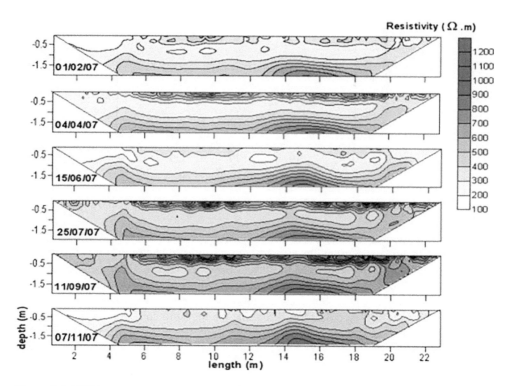

Figure 3.5 ERT (resistivity corrected at 25°C) during the year of 2007 (Brunet *et al.*, 2010)[2]

measured using TDR. Archie's law was calibrated in the laboratory to quantify moisture content and water deficit from ERT. A constant porosity of 0.42 and soil solution with resistivity of 22 Ohm-m were considered in the calibration. Based on the laboratory test results, the cementation (m) and saturation coefficients (n) were determined as 1.25 and 1.65, respectively. *In-situ* soil moisture content and water deficits were calculated from the calibrated Archie's law at 25°C temperature. The authors indicated that interpretation of water content or water deficit from resistivity is sensitive to temperature, water solution resistivity, porosity, and inversion algorithm of resistivity tests. The ERT profiles and comparisons of predicted and observed water content at different depths are presented in Figure 3.5 and Figure 3.6. The solid lines in Figure 3.6 indicate ERT-predicted moisture contents.

3.2.2 Unit weight

The study performed by Kibria and Hossain in 2012 indicated that resistivity is affected by soil unit weight. To determine the correlation of soil resistivity with unit weight, resistivity tests were conducted at different dry unit weight conditions, while keeping the gravimetric moisture content constant. Tests were conducted on four soil samples, sample 1, sample 2,

---

2 Reprinted from Journal of Hydrology, Vol. 380, Issue 1-2, Pascal Brunet, Rémi Clément, Christophe Bouvier, Monitoring soil water content and deficit using Electrical Resistivity Tomography (ERT) – A case study in the Cevennes area, France, 146-153, Copyright (2010), with permission from Elsevier.

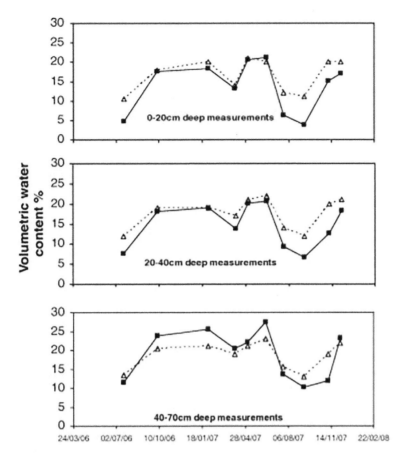

Figure 3.6 Comparison of ERT and TDR predicted water content at depths (a) 0–20 cm (b) 20–40 cm (c) 40–70 cm (Brunet *et al.*, 2010)[3]

sample 3, and sample 4 at moisture content of 18%, 24%, and 30% in each soil sample. The designations of the test specimens are presented in Table 3.1. The dry unit weight for each sample was varied from 11.8 kN/m³ to optimum for all soil samples. The variation of resistivity with unit weight at 18% moisture content is presented in Figure 3.7.

It was observed that soil resistivity decreased with the increase of unit weight in each condition. Soil resistivity decreased almost linearly, with an average rate of 1.91 Ohm-m/ (kN/m³), between moist unit weight of 13.92 to 15.72 kN/m³ at 18% moisture content. The changes in soil resistivity were significant up to moist unit weight of 15.72 kN/m³ in soil samples 1 and 3 at 18% moisture content. Soil resistivity decreased from 14.2 to 7.7 Ohm-m for the increase of moist unit weight from 13.92 to 15.72 kN/m³ in soil sample 3 at 18%

3 Reprinted from Journal of Hydrology, Vol. 380, Issue 1-2, Pascal Brunet, Rémi Clément, Christophe Bouvier, Monitoring soil water content and deficit using Electrical Resistivity Tomography (ERT) – A case study in the Cevennes area, France, 146-153, Copyright (2010), with permission from Elsevier.

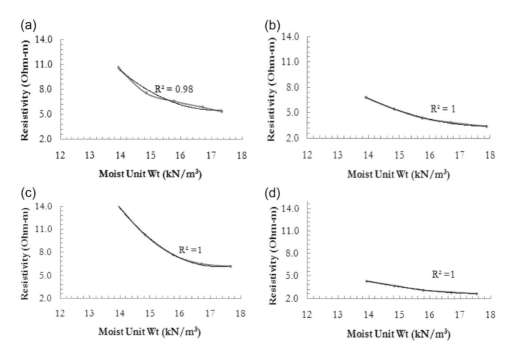

Figure 3.7 Variations of resistivity with moist unit weight at 18% moisture content: (a) Sample 1 (b) Sample 2 (c) Sample 3 (d) Sample 4 (Kibria and Hossain, 2012), with permission from ASCE

moisture content. The range of reduction in soil resistivity was from 10.7 to 6.6 Ohm-m in soil sample 1 at similar conditions. However, soil resistivity decreased with an average rate of 0.51 Ohm-m/(kN/m³) for further increases in moist unit weight in 18% moisture content in the soil samples. The observed variation in soil resistivity with unit weight was not substantial at 30% moisture content.

The variation of soil resistivity with unit weight can be explained by the study of Abu Hassanein *et al*. (1996). An increase in moist unit weight is associated with the increase in degree of saturation. More pronounced bridging occurs between the particles at a high degree of saturation. In addition, an increase of moist unit weight is associated with remolding of clay clods, elimination of interclod voids, and reorientation of particles (Abu Hassanein *et al*., 1996). Therefore, soil resistivity decreases with the increase of moist unit weight.

According to Mitchell and Soga (2005), reduction in the large pores and breakdown in flocculated open fabric occur during remolding of clay soil. As a result, the conduction path in soil is reduced at high unit weight. Test results showed that resistivity did not change significantly after 15.72 kN/m³. This might have been caused by the breakdown of flocculated fabric at high unit weight condition and associated reduction in current flow path.

The variations of resistivity values observed in the study of Kibria (2014) were plotted against dry unit weights at different moisture contents, as presented in Figure 3.8. It was

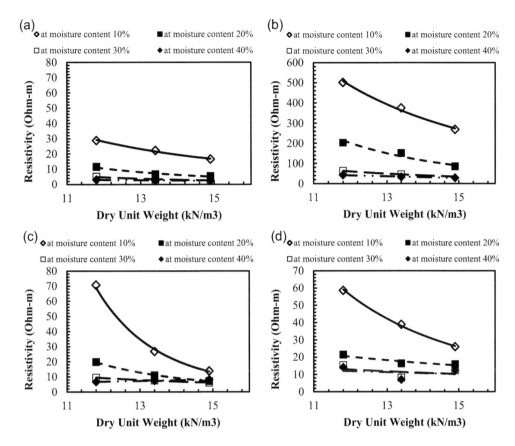

Figure 3.8 Variations of resistivity with dry unit weights for compacted clays at different moisture contents: (a) Ca-bentonite, (b) Kaolinite, (c) CL, and (d) CH

observed that resistivity reduction ranged from 5.5 to 12.6 Ohm-m in between 11.8 and 14.9 kN/m³ dry unit weights at 20% moisture content in Ca-bentonite, CL, and CH samples. However, resistivity reduced as much as 117.5 Ohm-m in kaolinite at this condition.

3.2.3 Degree of saturation

An increase of degree of saturation causes reduction in soil resistivity; however, the relationship is highly influenced by the critical degree of saturation. Critical degree of saturation corresponds to the minimum amount of water required for the development of a continuous water film around the soil particles. Typically, an abrupt increase of soil resistivity occurs below the critical degree of saturation (Bryson, 2005).

According to Rinaldi and Cuestas's (2002) study, the conductivity vs degree of saturation is concave upward, as presented in Figure 3.9. The authors indicated that the observed variations might occur due to the reduction in pore space and enhanced contacts between the particles at high degree of saturation.

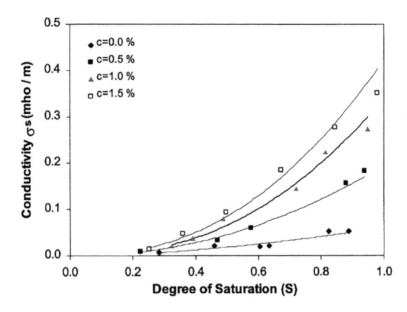

Figure 3.9 Influence of degree of saturation in soil conductivity (Rinaldi and Cuestas, 2002), with permission from ASCE

Matsui *et al.* (2000) performed a study on the correlation between electrical resistivity and physical properties of rock. Various types of granites and sedimentary rocks of Japan were utilized for the study. The rock specimens were saturated with tap water, and resistivity of the samples was measured at different stages of natural drying and artificial desiccation. The study results indicated that the resistivity decreased with the increase of degree of saturation up to a certain level; however, the variation was insignificant beyond that point.

Abu Hassanein *et al.* (1996) conducted resistivity measurements of four different soils at different initial degrees of saturation. It was observed that the electrical resistivity was inversely correlated with the initial degree of saturation. It was also noted that the initial degree of saturation and electrical resistivity were independent of compaction effort.

The variations of degrees of saturation with resistivity for fat clays obtained from Kibria and Hossain (2012) are presented in Figure 3.10. Sample designations are presented in Table 3.1. To obtain the degree of saturation, a specific gravity of 2.65 was assumed. It was observed that soil resistivity decreased with the increase of degree of saturation. Average soil resistivity of the samples was 6.7 Ohm-m at 40% degree of saturation; however, the average soil resistivity decreased to 3.2 Ohm-m at 90% degree of saturation.

Another important aspect of the test results was the variations of soil resistivity at similar degrees of saturation. Resistivity was 4.3 Ohm-m in sample 4 at 40% degree of saturation, which was low compared to other samples at that condition (Figure 3.9). The variations of soil resistivity at a specific condition might have been due to varied specific surface areas of the samples.

The variations of resistivity of the compacted samples with degrees of saturation obtained from Kibria and Hossain (2014) are illustrated in Figure 3.11. Test results showed that the degree of saturation significantly influences the electrical resistivity of soils. Electrical resistivity decreased as much as 11 times the initial value (28.6 to 2.6 Ohm-m) for a 23% to 100%

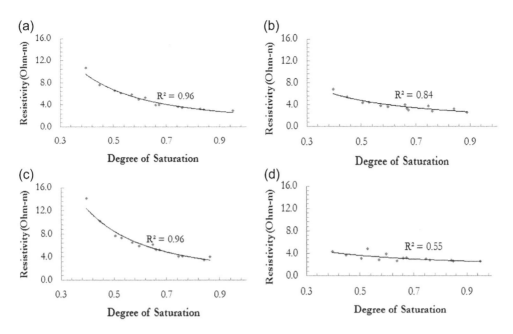

Figure 3.10 Variation of soil resistivity with the degree of saturation: (a) Sample 1 (b) Sample 2 (c) Sample 3 (d) Sample 4 (Kibria and Hossain, 2012), with permission from ASCE

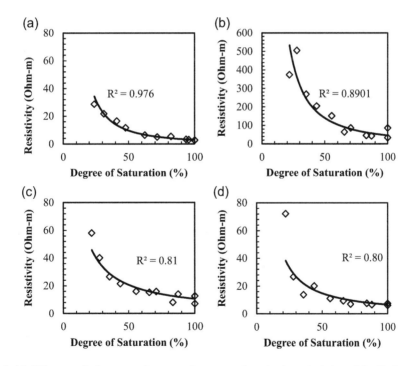

Figure 3.11 Effects of degree of saturation on electrical resistivity: (a) Ca-bentonite, (b) Kaolinite, (c) CL, and (d) CH (Kibria and Hossain, 2015)[4]

4 Reprinted from Waste Management, Vol. 39, Golam Kibria, Md. Sahadat Hossain, Investigation of degree of saturation in landfill liners using electrical resistivity imaging, 197-204, Copyright (2015), with permission from Elsevier

increase of degree of saturation in Ca-bentonite. The observed reductions ranged between 373 to 33 Ohm-m in kaolinite, 58.1 to 7 Ohm-m in CL, and 72 to 6 Ohm-m in CH at this condition.

The resistivity of unsaturated soil can be related to saturated soil by using the Keller and Frischnecht (1966) model, as follows:

$$\frac{R}{R_{100}} = S^{-B}$$

(3.3)

where, R and R_{100} are resistivity at unsaturated and saturated condition, respectively, S is the degree of saturation, and B is an empirical exponent.

Therefore, the observed results were normalized using resistivity at saturated condition to determine the exponent in the soil samples under consideration. It was observed that the exponents of Ca-bentonite, kaolinite, CL, and CH specimens were 1.72, 1.61, 0.96, and 1.15, respectively. The normalized resistivity vs. saturation curves are illustrated in Figure 3.12.

The electrical resistivity results were also plotted with degrees of saturation of undisturbed soils, as presented in Figure 3.13 (Kibria and Hossain, 2016). For an increase of saturation from 31% to 100%, resistivity decreased as much as sixteen fold (49.4 to 3.2) in the test results.

Based on the statistical analysis of the experimental results, the correlation between degree of saturation and resistivity was investigated for compacted and undistributed clayey soils by Kibria (2014), as presented in Figure 3.14 and Figure 3.15.

3.2.4 Volumetric moisture content

Kalinski and Kelly (1993) conducted a laboratory investigation to determine volumetric moisture content from electrical conductivity of soil. The electrical resistivity of soil was measured, using a four-probe circular cell. Porous plates were utilized to extract water from the soil, and electrical conductivity (EC_w) was determined. The experimental results indicated

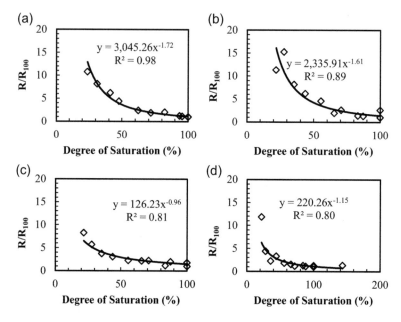

Figure 3.12 Normalized resistivity vs. degree of saturation: (a) Ca-bentonite, (b) Kaolinite, (c) CL, and (d) CH

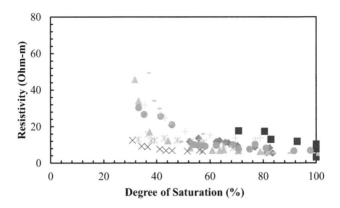

Figure 3.13 Electrical resistivity variations with degree of saturations in undisturbed soils (Kibria and Hossain, 2016), with permission from ASCE

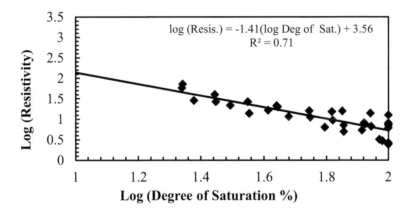

Figure 3.14 Development of correlation between electrical resistivity and degree of saturation for Ca-bentonite, CH, and CL

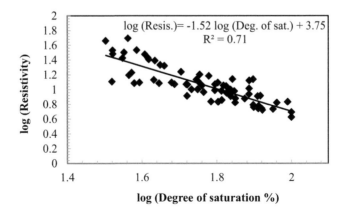

Figure 3.15 Development of correlation between electrical resistivity and degree of saturation of undisturbed soil samples

that the EC_o/EC_w (ratio of soil conductivity and pore water conductivity) increased with the increase of volumetric water content, as presented in Figure 3.16. In addition, the following regression equation was developed to determine volumetric water content, assuming surface conductivity of 0.24 mho/cm (ECs = 0.24 mho/ cm). Predicted and measured volumetric moisture contents were in good agreement.

$$EC_0 = EC_s + EC_w\theta(1.04\theta - 0.09) \qquad (3.4)$$

Variations of soil resistivity with volumetric moisture content obtained from the study performed by Kibria and Hossain (2012) are presented in Figure 3.17.

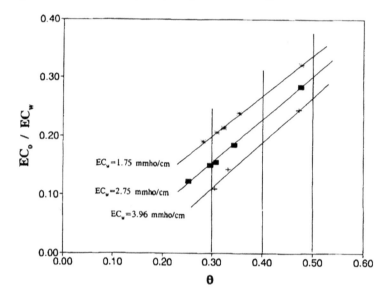

Figure 3.16 Relationship between the ratio of bulk soil to pore water conductivity with volumetric moisture content (Kalinski and Kelly, 1993)[5]

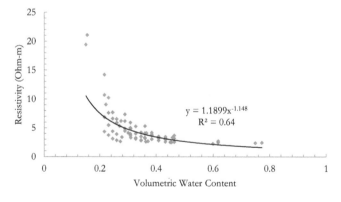

Figure 3.17 Variation of soil resistivity with volumetric water content (Kibria and Hossain, 2012), with permission from ASCE

5 Reproduced, with permission from Kalinski and Kelly, 1993, copyright ASTM International, 100 Barr Harbor Drive, West Conshohocken, PA 19428.

From the relationships of volumetric water content with dry unit weight of soil and unit weight of water and gravimetric water content, the following equation was developed:

$$\rho = 136.89 \left[\gamma_{m.} . w / (1+w) \right]^{-1.148} \tag{3.5}$$

where, ρ = Resistivity in Ohm-m, γ_m = Moist unit weight in pcf, w = gravimetric moisture content.

The three-dimensional surface area was obtained from the equation, using Mathematica software. The obtained surface is presented in Figure 3.18.

It is evident from Figure 3.18 that the soil resistivity decreased with the increase of both moisture content and unit weight. At moisture content beyond 40%, the surface became almost parallel to the plane of moisture content and unit weight.

The variation of resistivity with volumetric water content in compacted clay samples observed in the study of Kibria (2014) is presented in Figure 3.19. An increase in volumetric water content from 12% to 61% caused 11.8 (28.6 to 2.4 Ohm-m), 12.9 (373 to 29 Ohm-m), 4.6 (58.1 to 12.6 Ohm-m), and 9 (72.2 to 8 Ohm-m) times reductions in resistivity of Ca-bentonite, kaolinite, CL, and CH samples, respectively.

The variations of electrical resistivity with volumetric moisture content of undisturbed soil are presented in Figure 3.20. It was determined that resistivity decreased from 49.4 to 4.2 Ohm-m for the increase of volumetric moisture contents from 14.1% to 40.7% in the B2–10 sample. Moreover, significant variations were identified between volumetric moisture contents 14% to 31%, with the total reduction in resistivity being as much as 44 Ohm-m in

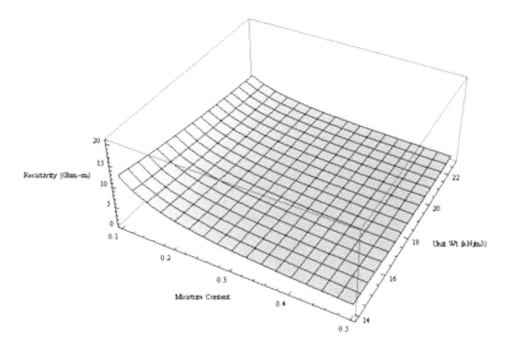

Figure 3.18 Three-dimensional surface area, combining moist unit weight and moisture content with resistivity (Kibria and Hossain, 2012), with permission from ASCE

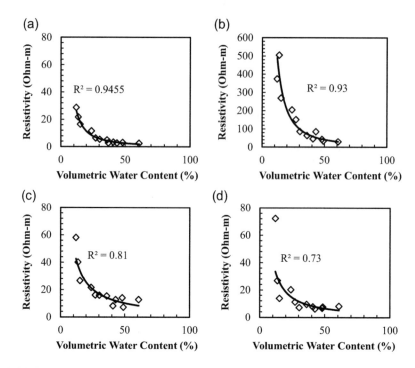

Figure 3.19 Resistivity variations with volumetric water contents in compacted clays: (a) Ca-bentonite, (b) Kaolinite, (c) CL, and (d) CH

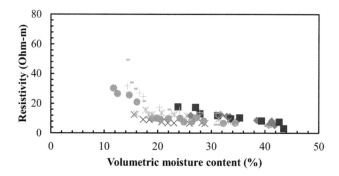

Figure 3.20 Resistivity variations with volumetric water content in undisturbed soils

these moisture ranges. In addition, the resistivity observed at a specific moisture content was different for different soil samples. At 20% volumetric moisture content, the resistivity was 17 Ohm-m in the B2–5 sample. However, the resistivity values were 16, 12, and 9 Ohm-m at this moisture content in B2–10, B2–20, and B2–25 specimens, respectively. The variations of resistivity at specific moisture contents might occur due to the varied CEC and PI of the samples.

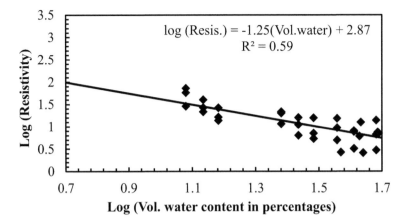

Figure 3.21 Development of correlation between electrical resistivity and volumetric moisture content for compacted Ca-bentonite, CH, and CL

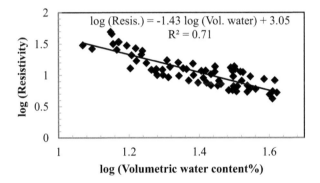

Figure 3.22 Development of correlation between electrical resistivity and volumetric content of undisturbed soil samples

The developed correlations between electrical resistivity and volumetric moisture content of disturbed and undisturbed soil samples are presented in Figure 3.21 and Figure 3.22. The developed correlation for compacted soils was also compared with the previous results.

3.2.5 Compaction condition

Rinaldi and Cuestas (2002) performed a laboratory investigation to evaluate the relationship between electrical conductivity and compaction. The soil samples were put through a No. 40 sieve and were compacted at 18% moisture content. Compaction was conducted using the Standard Proctor method in a rectangular mixing pan. After compaction, conductivity was measured, using a four-probe electrode device. Based on the experimental results, the iso-conductivity contour was obtained from the test, as illustrated in Figure 3.23.

According to Figure 3.23, the conductivity in the central portion was higher than on the right-hand side and border. The authors indicated that the variation of conductivity was

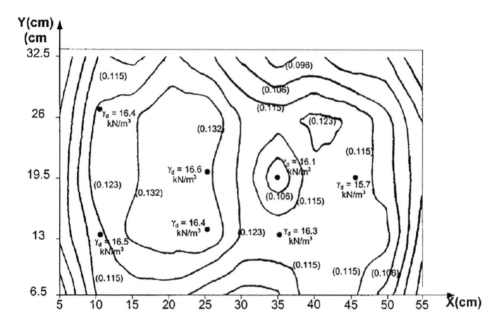

Figure 3.23 Iso-conductivity contour of compacted sample (parentheses show electrical conductivity in mho/m.) (Rinaldi and Cuestas, 2002), with permission from ASCE

attributed to the variation of soil unit weight. The unit weight was higher on the left-hand side and decreased on the right-hand side and border due to the low stiffness of the wall of the mixing pan.

McCarter (1984) conducted a study to evaluate the effect of the air void ratio in soil resistivity on Cheshire and London clay. A substantial decrease in soil resistivity was observed for the increase of degree of compaction or degree of saturation. The study results emphasized that the compaction condition and the moisture content are important factors in resistivity variations.

Abu Hassanein *et al.* (1996) performed a comprehensive study on the effects of molding water content and compacting efforts in soil resistivity. The soil specimens were compacted, using three different compaction methods: (1) Standard, (2) Modified, and (3) Reduced Proctor. It was observed that the resistivity was high when the soil was compacted at dry optimum, and it was low when compacted at wet optimum. Moreover, resistivity was sensitive to molding water content below optimum condition. At wet-of-optimum, resistivity was almost independent of molding water content. The authors indicated that this correlation might be useful in evaluating the compaction condition of soil. The test results are presented in Figure 3.24.

Kibria (2011) conducted a study to evaluate the compaction condition of fat clay specimens, using electrical resistivity. The optimum moisture contents and dry unit weights of fat clay samples are presented in Table 3.2.

Soil resistivity tests were conducted at moisture content and dry unit weight, corresponding to the compaction curve in each sample. Test results showed that resistivity was high

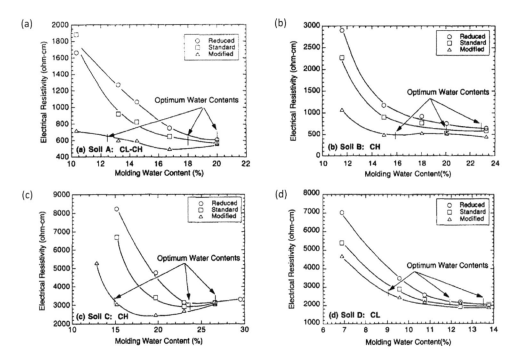

Figure 3.24 Relationship among electrical resistivity, molding water content, and compaction effort for different soils: (a) Soil A (b) Soil B (c) Soil C (d) Soil D (Abu Hassanein *et al.*, 1996), with permission from ASCE

Table 3.2 Optimum moisture content and dry unit weight of soil samples

Designation of Samples	Optimum Moisture Content (%)	Dry Unit Weight (kN/m³) (kN/m³)
SAMPLE 1	23.0	14.7
SAMPLE 2	21.5	15.2
SAMPLE 3	24.6	15.0
SAMPLE 4	24.0	14.9

when samples were compacted at dry of optimum. Resistivity decreased significantly with the increase of moisture content and unit weight. At the wet side, soil resistivity was low. The schematic of the test results is presented in Figure 3.25.

The average reduction in soil resistivity ranged from 4.75 to 3.4 Ohm-m when the soil samples were compacted at dry of optimum. However, the observed soil resistivity was as low as 1.8 Ohm-m (Soil Sample 4) at wet-of-optimum. The average range of resistivity was 2.2 to 2.6 Ohm-m at the wet side of the compaction curve. Therefore, soil resistivity was almost independent of molding water content and dry unit weight at the wet side of the compaction curve.

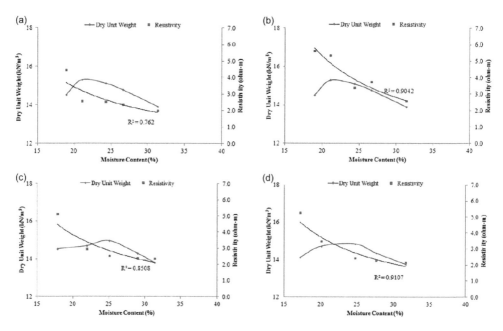

Figure 3.25 Variations of soil resistivity at different compaction conditions of the soil samples: (a) Sample 2 (b) Sample 3 (c) Sample 4 (d) Sample 1

The variations of soil resistivity with compaction condition can be discussed by looking at the structural changes of soil during compaction. Clay clods are difficult to remold, and interclod pores are large when compacted at dry of optimum (Daniel and Benson, 1990). The pores are generally filled with dielectric air at this condition. The contacts between the particles are poor because of the presence of distinct clods at low dry unit weight. Therefore, the observed resistivity might be high at dry of optimum due to the presence of air-filled voids and poor particle-to-particle contact.

Variations in resistivity with molding water content can occur due to structural changes of soil during compaction. At low compaction effort and dry of optimum water content, clay clods are difficult to remold. The interclod pores are also relatively large pores that are filled with dielectric air, diffuse double layers are not fully developed, and inter-particle contacts are poor at this condition. In contrast, clods of clay can be easily remolded at wet-of-optimum and high-compaction effort, which results in an increase in saturation. An enhanced particle-to-particle contact and formation of a bridge between the particles improves the electrical conductivity of soil (Abu Hassanein *et al.*, 1996).

3.2.6 Pore water characteristics

Electrical conductivity in a porous medium depends on the mobility of ions present in the pore fluid. The hydration of precipitated salts leads to the formation of electrolytes in the pore water of clayey soil. Hydrated cations and anions move towards cathode and anode under the applied electric field. The movement of ions reaches a terminal velocity when applied electric field and charge interactions (viscous drag force) are in equilibrium (Santamarina *et al.*, 2001). Ion mobility is defined as the terminal velocity of an ion subjected to a unit electric field and can be presented by the Einsten-Nernst equation below:

$$u = \frac{V(ion)}{E} = \frac{ze}{6 \; \eta R_h} \tag{3.6}$$

Here, V(ion) is the velocity of ion (m/s), E is the electric field (V/m), z is the valence of ion, η is the viscosity of the solution (Pa.s), e is the charge of electron (1.602×10–19 C) and Rh is the Stokes' radius of the hydrated ions.

The electrical conductivity is affected by ions present in the soil because of the varied ionic mobility. Ions such as H^+, OH^-, SO_4^{2-}, Na^+, and Cl^- are present in the soil, but they do not affect conductivity in the same way because of the differences in their ion mobility. A study conducted by Kalinski and Kelly (1993) indicated that electrical resistivity of soil decreases with the increase of pore water conductivity. Based on the experimental results, the following equation was developed to estimate pore water conductivity:

$$ECw = ECo - \frac{ECs}{\theta(a\theta + b)} \tag{3.7}$$

where, ECw = pore water electrical conductivity, ECs = apparent soil particle surface electrical conductivity, Eco = bulk soil electrical conductivity, Θ = volumetric water content, and a and b = constant.

Rinaldi and Cuestas (2002) emphasized the influence of sodium chloride and other electrolytes on the loess soil of Argentina. The soil samples were compacted at a constant density, and electrolytes were added. Electrical conductivity was measured at different concentrations of various electrolytes. The study results suggested a linear relationship between conductivity of soil and electrolytes, as presented in Figure 3.26. According to the study, samples

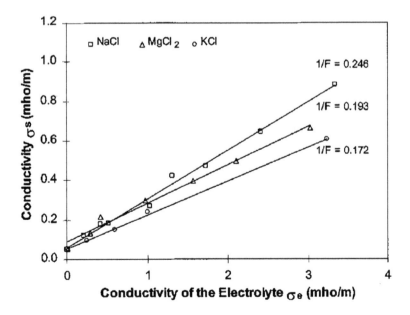

Figure 3.26 Relationship between conductivity of saturated samples at different electrolytes (Rinaldi and Cuestas, 2002), with permission from ASCE

containing sodium showed the highest conductivity, followed by magnesium and potassium. The difference in conductivity was due to the ion mobility of different electrolytes, adsorption, and soil structure.

Although electrical conductivity increases with the presence of ions in pore water, the mobility of the charge may be restricted due to the availability of ions at a high concentration. A reduction in conductivity may occur at that condition (Santamarina *et al.*, 2001).

Studies were conducted by Kibria (2014) and Kibria and Hossain (2017) to determine the effects of pore water conductivity and sulphate ions on the electrical resistivity of soils. The sulphate ion contents of the pore water were investigated using the ion chromatography (IC) method. Four types of specimens, Ca-bentonite, kaolinite, CH, and CL, were utilized in the pore water extraction. The soil specimens and clay minerals were thoroughly mixed with deionized water at their liquid limits using a rotary mixer, and the slurry was transferred to a modified consolidometer. The consolidometer was equipped with a connection pipe beneath the porous stone, and pore water was extracted from soil specimens under applied loads at different stages of consolidations. The collected pore water was transferred to the test tube and stored at a temperature of 4.0°C. The electrical conductivity of the pore water was measured using a bench-top conductivity/pH meter. Extracted pore water was injected into the sodium bicarbonate solution. The compound mixture was then allowed to pass through the separation columns. The composed ions were separated based on the interaction of dissolved ions in the pore water with the carrier fluid and adsorbent. A conductivity detector was used to determine the concentration of cations and anions. In addition, an anionic suppressor was utilized to reduce the movement of carrier fluid and enhance the conductance of separated ions. The study results are presented in Table 3.3.

It was observed that the electrical conductivity of the pore water of Ca-bentonite, kaolinite, CL, and CH was 1805, 414, 1028, and 1436 microsiemens/cm, respectively. The electrical conductivity of the pore water extracted from Ca-bentonite was high compared to other samples, indicating the presence of a large amount of surface charge. IC results showed that the concentration of sulphate ions varied significantly among the pore water of the soil samples. The concentration of sulphate ions of Ca-bentonite and kaolinite were 824.3 and 88.5 mg/L, respectively.

The resistivity responses at varied pore water conductivity obtained from Kibria (2014) are illustrated in Figure 3.27. The pore water conductivity significantly influenced the electrical resistivity of the soil samples. According to the results, the pore water conductivity of kaolinite was 414 microsiemens/cm, and associated resistivity values were 435, 140, 75, and 45 Ohm-m at 25, 50, 75, and 100% saturation, respectively. The pore water conductivity of Ca-bentonite, CL, and CH were 1805, 1028, and 1436 micro-Siemens/cm, respectively, and

Table 3.3 Observed electrical conductivity and sulphate ions from IC

Sample	Electrical Conductivity (micro siemens/cm)	Sulphate (mg/L)
Ca– bentonite	1805	824.31
Kaolinite	414	88.45
CH	1028	382.52
CL	1436	197.77

Note: Electrical conductivity and sulphate results of Ca-bentonite and kaolinite were obtained from Kibria and Hossain (2017)

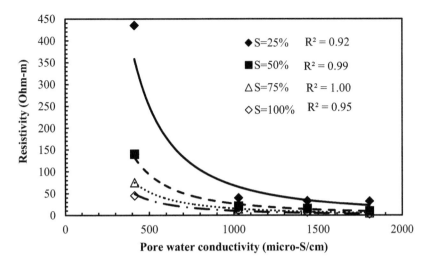

Figure 3.27 Variations of resistivity of samples with pore water conductivity

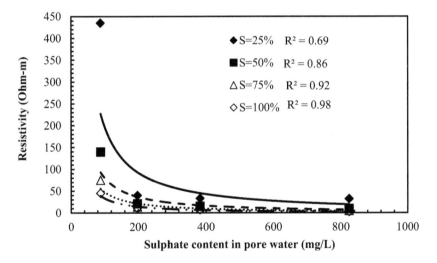

Figure 3.28 Variation of resistivity of samples with sulphate content in pore water

the observed resistivity was below 50 Ohm-m at each saturation level due to the high conductivity of the extracted pore water.

Ion chromatography results of extracted pore water showed significant variations of sulphate concentrations among the samples. Therefore, sulphate ions of the pore water were correlated with resistivity at different degrees of saturation. Figure 3.28 indicates that an increase of sulphate ions results in substantial reduction in resistivity. Resistivity reduced from 435 to 32 Ohm-m for the increase of sulphate concentrations from 88.5 to 878.7 mg/L at 25% degree of saturation. The observed variations might have occurred due to the enhanced ionic conduction at high sulphate contents.

3.2.7 Ion composition and minerology

The electrical resistivity of soils is affected by constituent ion composition and mineralolgy. The study conducted by Kibria and Hossain (2012) indicates that the ion composition of test specimens influences resistivity. The percentages of Mg, Mn, Fe, Ti, Al, Si, S, P, and K ion composition in the test specimens were not substantial (<2%) in the soil samples; however, calcium content of the samples ranged from 7% to 13.9%. The changes in resistivity with calcium contents are presented in Figure 3.29. Tests were conducted at moisture contents of 18%, 24%, and 30% in order to determine the effects of ion composition on soil resistivity. Dry unit weights of the samples were 11.8 and 13.4 kN/m³ during the tests. Test results showed that soil resistivity increased from 4.3 to 14.2 Ohm-m for an increase of calcium ion from 8.3% to 13.9% at 18% moisture content and 11.8 kN/m³ dry unit weight. The rate of increase of soil resistivity decreased with the increase of moisture content. At 30% moisture content and 11.8 kN/m3 dry unit weight, the soil resistivity was between 3.2 and 5.3 Ohm-m.

The variations might have occurred because a water film didn't form around large areas of soils with low water content and the ionic conductor calcium couldn't penetrate the solutions with low moisture content. Therefore, the surface area controlled the soil resistivity at this condition. With the increase of moisture content, the calcium ion might become mobile, decreasing the soil resistivity. The soil resistivity was almost constant and independent of unit weight and calcium content at 30% moisture content, suggesting that the soil resistivity was highly sensitive to the moisture content and specific surface area.

Another study conducted by Kibria and Hossain (2014) showed that the resistivity varied with the bentonite content. The objective of that study was to determine the variations in electrical resistivity at different bentonite contents. Na-bentonite and Ca-bentonite were mixed with fine sands to identify their effects on resistivity. Figure 3.30 shows that the resistivity decreased from 14.4 to 8.8 Ohm-m with an increase in Na-bentonite from 20% to 100% at 40% degree of saturation. Resistivity decreased from 29.2 to 10.1 Ohm-m with a 20% to 100% increase of Ca-bentonite at this saturation.

Abu Hassanein et al. (1996) described an innovative method to determine the bentonite content of soils using electrical conductivity. A detail sedimentation analysis was performed on the test specimens, and a calibration curve was developed to correlate electrical conductivity with bentonite concentration at a given temperature. The proposed methods were simple

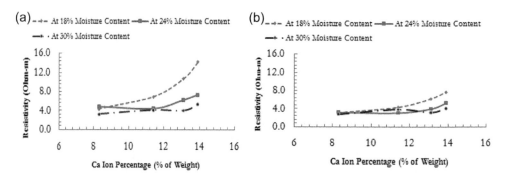

Figure 3.29 Percentage of calcium ions and their variations with resistivity at dry unit weight (a) 11.8 kN/m³ (b) 13.4 kN/m³ (Kibria and Hossain, 2012), with permission from ASCE

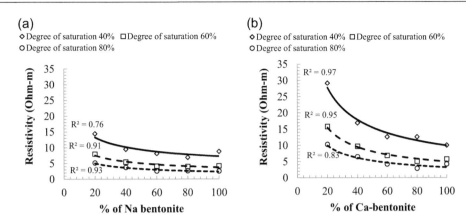

(a) Na-bentonite
(b) Ca-bentonite

Figure 3.30 Variation of resistivity with bentonite percentages (Kibria and Hossain, 2014), with permission from ASCE

and required less time to measure the bentonite content; however, the performance of the method was not evaluated in the field condition.

3.2.8 Structure, packing, and hydraulic conductivity

Zha *et al.* (2007) presented a study on the evaluation of expansive soil using electrical resistivity measurement. The average formation factor, shape factor, and electrical anisotropy index were investigated at different stages of swelling. It was observed that the formation factor and shape factor varied linearly with the increase of swell percentages. In addition, the initial, primary, and secondary swellings were determined using the relationship among log-time, formation factor, and shape factor. According to the authors, the decrease in the average formation factor was related to microstructure changes; formation destruction; increase in water content and porosity; and decrease in strength, cementation, and stability of soils.

Based on the available literaure, it is evident that electrical resistivity depends on porosity, structure, saturation, and tortuosity of soil. As hydraulic conductivity also depends on these parameters, several researches have attempted to correlate hydraulic conductivity with electrical resistivity (Bryson, 2005). Sadek (1993) performed a study to explore the possibility of using electrical conductivity as an alternative to hydraulic conductivity in compacted clay liners. An extensive research program was developed which included (1) comprehensive review of the influential parameters affecting electrical and hydraulic conductivity, (2) development of a theoretical model to incorporate pore water conductivity and surface conductance, and (3) design of new equipment to study the electro-kinetic properties. The study results indicated that the electrical conductivity of soils was not sensitive enough for use as an alternative to hydraulic conductivity. The electrical conductivity of a sample with dispersed structure and low hydraulic conductivity was similar to the sample with flocculated structure and high hydraulic conductivity. The author indicated that incorporating surface conductance and internal pore geometry using the "Cluster model" might provide a better correlation; however, this required quantification of internal geometry. Although, the electrical

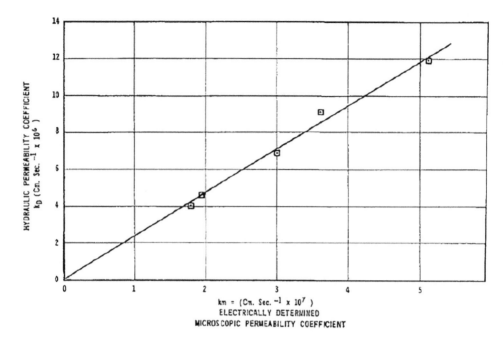

Figure 3.31 Correlation between microscopic and hydraulic permeability coefficients during consolidation of a kaolinite-hydrite MP made homo-ionic to 0.001N NaCl (Arulanandan, 1969)[6]

conductivity was not proven as a reliable indicator of hydraulic conductivity, the study provided useful insight to the electrical properties of soils.

Abu Hassainein *et al.* (1996) considered four soil specimens to correlate hydraulic conductivity and resistivity; however, the study results did not indicate a specific correlation between these parameters.

Arulanandan (1969) discussed the electrical conductivity of saturated kaolinite, illite, and montmorillonite clay in the 50–10[8] cycle/sec. frequency range. It was observed that the electrical conductivity increased with an increase of frequency, and the required frequency for the first dispersion was independent of particle size. The microscopic permeability coefficients were evaluated based on the electrical properties. The study results indicated that the coefficients of Darcy's law were correlated with several electrical properties, i.e., conductivity of the AC and DC range and ratio of phenomena logic transport coefficient. A schematic of the study results is presented in Figure 3.31.

3.2.9 Cation Exchange Capacity (CEC) and Specific Surface Area (SSA)

Adsorbed cations are significant in electrical resistivity of medium and fine-grained soil. It is evident from the literature that the physico-chemical properties such as adsorbed ions,

---

6 From Arulanandan, 1969. Reproduced with kind permission of The Clay Minerals Society, publisher of Clays and Clay Minerals.

pore water conductivity, and surface charge are correlated with the cation exchange capacity (CEC) of the soils (Friedman, 2005; Tabbagh and Cozenza, 2007; Schwartz *et al.*, 2008).

Kibria and Hossain (2012) performed a study to determine the effect of a specific surface area (SSA) in soil resistivity, as presented in Figure 3.32. Figure 3.32 indicates that the soil resistivity increased with the increase of the specific surface area. The increments were more pronounced at 18% moisture content. Soil resistivity increased from 4.3 to 14.2 Ohm-m with the increase of surface area from 69.6 to 107.1 m²/gm at 18% moisture content and 11.8 kN/m³ dry unit weight, and increased from 3.2 to 5.3 Ohm-m at 30% moisture content and 11.8 kN/m³ dry unit weight. Soil resistivity ranged from 2.8 to 3.2 Ohm-m at 30% moisture content and 14.2 kN/m³ dry unit weight in all of the soil samples.

With an increase of surface area, more water is required for the formation of a water film around the fine particles. In the absence of water film, bridging cannot occur between the soil particles and ionic conduction does not take place. Therefore, the lack of formation of water film around the particles due to a large specific surface area might have caused the observed variation at 18% water content. With the increase of moisture content, water bridging between the particles occurred. Therefore, the rate of variation was 0.08 Ohm-m/ (m²/gm) at 24% moisture content and 0.04 Ohm-m/ (m²/gm) in 30% moisture content at 11.8 kN/m³ dry unit weight. At 30% moisture content and 14.2 kN/m³ dry unit weight, soil resistivity was independent of the surface area because of the formation of a water film around particles in a compacted soil condition.

The resistivity responses at varied CECs for compacted clay samples obtained from Kibria and Hossain (2014) were plotted and are depicted in Figure 3.33. It was observed that the resistivity results were different for the specimens at specific degrees of saturation and

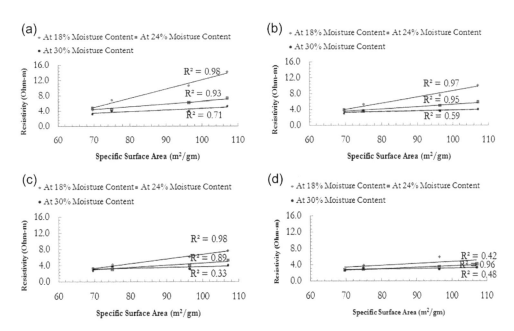

Figure 3.32 Variation of soil resistivity with specific surface area at dry unit weight: (a) 11.8 kN/m³ (b) 12.6 kN/m³ (c) 13.4 kN/m³ (d) 14.2 kN/m³ (Kibria and Hossain, 2012), with permission from ASCE

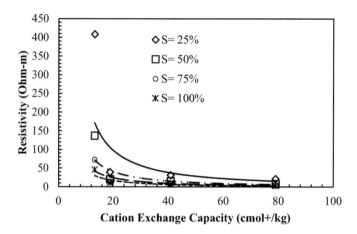

Figure 3.33 Resistivity variations with CEC of compacted clays (Kibria and Hossain, 2015)[7]

temperatures. The variations might have occurred due to varied isomorphous substitutions of clay particles. The experimental scatter plots indicated that the coefficients of regression were the highest in power-fitted trend lines. Therefore, power functions were utilized to fit the experimental results.

The resistivity was significantly affected by CEC at relatively low degrees of saturation. A total reduction in resistivity was as much as 385 Ohm-m for the increase of CEC from 13.3 to 79 cmol+/kg at 25% degree of saturation. The total reduction in resistivity was 43.3 Ohm-m at 100% degree of saturation in the above CEC range.

The electrical conductivity of soil increases with the increase of moveable ions in the pore water; consequently, soil resistivity decreases with the increase of CEC because of the presence of conductive ions. Moreover, kaolinite consists of exchangeable ions at the edges of the crystal, whereas bentonite is composed of exchangeable ions, both at the edges and within the lattice structure. Therefore, the contribution of precipitated ions is more significant in bentonite than in kaolinite (Holeman, 1970). According to Mitchell and Soga (2005), cation exchange reactions generally do not induce structural changes in clays. However, significant variations in physical and physicochemical properties may occur due to the change in CEC. Test results indicated that the CEC of different artificial soils significantly influenced resistivity at low degrees of saturation. Nonetheless, resistivity was less sensitive to CEC at a high degree of saturation (i.e. S = 100%).

The rate of reduction of electrical resistivity decreased with the increase of CEC. At a specific degree of saturation, the mobility of precipitated ions might decrease with the increase of CEC. Therefore, the response curve was flatter at high CEC than at the low ion exchange condition.

The CEC of artificial soil samples was plotted against resistivity, as illustrated in Figure 3.34. According to Figure 3.34, an increase in CEC caused a significant reduction in resistivity at 25% degree of saturation. For Ca-bentonite-sand specimens, resistivity decreased from 64 to 37 Ohm-m with the increase of CEC from 27.8 and 63.5 cmol+/kg, respectively

7 Reprinted from Waste Management, Vol. 39, Golam Kibria, Md. Sahadat Hossain, Investigation of degree of saturation in landfill liners using electrical resistivity imaging, 197-204, Copyright (2015), with permission from Elsevier

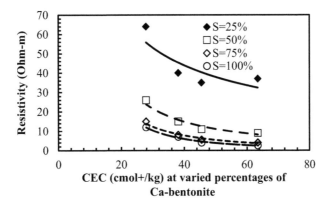

Figure 3.34 Resistivity variations with CEC of artificial samples (Kibria and Hossain, 2014), with permission from ASCE.

at this saturation. In contrast, resistivity reduced from 12 to 2.5 Ohm-m in this CEC range at 100% saturation. Therefore, test results emphasized the effect of ion exchange at low degrees of saturation.

3.2.10 Temperature

Electrical resistivity decreases with the increase of temperature because of the agitation of ions. It was observed that an increase in temperature per degree Celsius decreases electrical resistivity by 2.02% (Campbell *et al.*, 1948). According to the study of Abu Hassanein *et al.* (1996), an exponential relationship exists between electrical resistivity and soil above 0°C. As the electrical resistivity of soils depends on the temperature, experimental measurements should be corrected with respect to a reference temperature to explain the variations of moisture content, unit weight, soil structure, and soil type with resistivity.

Several conversion models were developed to express electrical resistivity at a reference temperature. Besson *et al.* (2008) performed a study to analyze the soil resistivity of two types of soil models at varied temperature conditions. Their study results suggested that the models were mostly empirical, and the parameters of the available models depended on the soil solution properties. The available models indicated good accuracy at high volumetric moisture contents; however, a relationship between resistivity and temperature was required to identify the best conversion model. Based on the investigation results, the following model was developed:

$$f\left(T_m, T_{ref}\right) = \left(\frac{T_{ref}}{T_m}\right)^s or \rho_{ref} = \rho_m \left(\frac{T_{ref}}{T_m}\right)^{-s} \tag{3.8}$$

Here, T_m = medium temperature, T_{ref} = reference temperature, ρ_{ref} = corrected resistivity at reference temperature, ρ_m = resistivity of the medium at T_m, and s = empirical parameter.

3.2.11 Consolidation properties

Consolidation in soil is associated with the dissipation of pore water, reduction in void ratio, and change in fabric morphology. Therefore, consolidation properties of soil can be evaluated by using electrical properties. McCarter and Desmazes (1997) investigated the changes in electrical conductivity of clayey soil in response to consolidation stages. A modified consolidation cell was utilized to measure the variations in electrical properties with void ratio at saturated condition. The electrical conductivity was determined in the vertical and horizontal directions. It was observed that the changes in void ratio and conductivity with effective stress were very similar. According to the authors, conduction in saturated soil occurred through continuous interstitial water. Therefore, the fractional volume of water and composition of pore fluid influenced electrical properties significantly. Nonetheless, electrical conductivity of soil decreased with the progression of consolidation process due to the dissipation of pore water.

Bryson (2005) correlated void ratio with conductivity from the curve obtained by McCarter and Desmazes (1997). The developed correlation provided a mean to differentiate consolidation properties from electrical conductivity of soils. A one-dimensional settlement equation and the compression index for normally consolidated clay are presented below (Bryson, 2005):

$$S = \frac{\Delta e}{(1+e)} H = \frac{\Delta \sigma}{1+\sigma v}(\xi) H \tag{3.9}$$

$$C_c = \varepsilon(\Delta \sigma) \log\left(\frac{P}{P_o}\right) \tag{3.10}$$

where, Δe = change in void ratio, e = initial void ratio, $\Delta \sigma$ = change in vertical conductivity, σ_v = Initial vertical conductivity, ξ = factor relating vertical conductivity and void ratio, H = sample height, P = consolidating pressure, and P_o = initial pressure.

McCarter et al. (2005) used electrical resistivity measurements to evaluate structural changes during consolidation. A typical oedometer was modified to simultaneously measure the load deformation and electrical conductivity of a specimen. The modified oedometer was able to accommodate a sample 185 mm high with a 250 mm diameter. The horizontal and vertical resistivity results were determined during loading and unloading conditions. Changes in soil structure, formation factor, and anisotropy were investigated. Based on the experimental results, Comina et al. (2008) developed an advanced EIT oedometer for the evaluation of 3D electrical tomography and the measurement of seismic wave velocity of soil specimens. The inner walls of the test cell were covered with stiff polyamide to ensure insulation during current flow and reduce wall friction. The concentric rings, with a permeability of 6×10^{-6} m/sec., were utilized for the drainage. The electrical resistivity was measured, using 42 electrodes hosted on the internal boundary of the cell. In addition, P- and S-wave velocities were determined using the sensors at the top and bottom caps of the cell. A vertical load was applied on the specimen under one-dimensional conditions. The diameter of the specimen was 130 mm and the height ranged between 20 and 60 mm. The newly-designed EIT oedometer, wave sensors, and measurement setup are illustrated in Figure 3.35.

A number of trial experiments were performed to assess the 3D imaging performance of the cell, using a homogenous sample and a sample with resistive and conductive inclusion. Ticino sand was utilized for the simultaneous measurement of deformation, shear wave, compression wave, and topographic imaging. It was observed that the P-wave velocity of the saturated sample did not vary significantly, while the S-wave velocity increased from 120 to 180 m/sec in the 100 to 400 kPa pressure range. The experimental results of electrical conductivity, void, and pressure are presented in Figure 3.36. Based on the preliminary results, the authors indicated that the newly designed cell was able to evaluate the transient

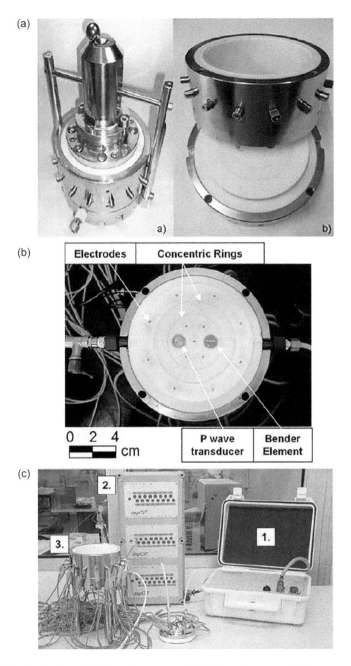

Figure 3.35 Newly-designed EIT oedometer: (a) EIT oedometer, (b) wave sensors, and (c) measurement setup (Comina *et al.*, 2008)[8]

8 Reproduced, with permission from Comina et al., 2008, copyright ASTM International, 100 Barr Harbor Drive, West Conshohocken, PA 19428

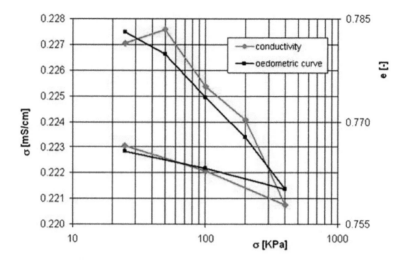

Figure 3.36 Experimental results of electrical conductivity, void, and pressure (Comina *et al.*, 2008)[9]

process of chemical diffusion and the preferential flow path during conduction and changes in mechanical properties.

3.2.12 Void ratio

Kim *et al.* (2011) conducted a study to determine the void ratio from resistivity in a seashore soil. For the purpose of the study, electrical resistivity was measured using a newly designed electrical resistivity cone probe (ERCP). The research was conducted, using two approaches. First, laboratory tests were conducted to obtain the well-known Archie's law, and then the results were validated with field data. Pore water was extracted from the soil by using a miniature centrifuge to calibrate Archie's law for the specific soil samples in the laboratory. After calibration, the porosity profile was obtained by using the coefficient of cementation (m parameter of Archie's law), which ranged from 1.4 to 2.0. To validate the results, two field tests were carried out in Incheon and Busan, Korea. In the field, ERCP was pushed at penetration rates of 1 mm/s and 3.3 mm/s into the sites. Undisturbed samples were obtained, using thin-walled samplers to estimate volume-based void ratio in the laboratory. Volume-based void ratio matched well with resistivity-based void ratio obtained from the calibrated Archie's law. According to the authors, the void ratio can be determined from resistivity when Archie's law is calibrated for a specific soil sample.

Kibria (2014) conducted a study to determine the relationship of void ratio with electrical resistivity. The void ratio of compacted clay specimens was plotted against resistivity at different moisture contents, as presented in Figure 3.37. Electrical resistivity increased from 16.5 to 28.8 Ohm-m for an increase of void ratio 0.59 to 1.01 at 10% moisture content in the Ca-bentonite specimen. However, the variation in resistivity was not significant at 40% moisture

9 Reproduced, with permission from Comina et al., 2008, copyright ASTM International, 100 Barr Harbor Drive, West Conshohocken, PA 19428

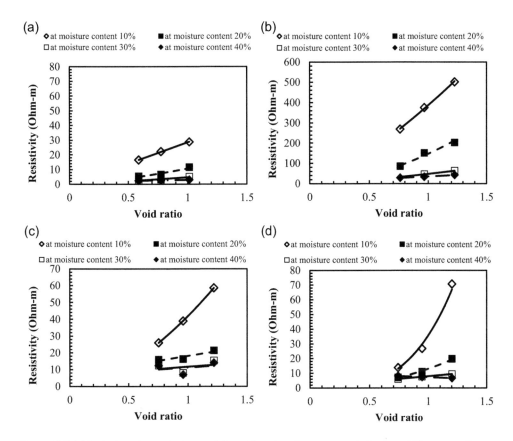

Figure 3.37 Variations in resistivity with void ratio of compacted clays at different moisture contents: (a) Ca-bentonite (b) Kaolinite (c) CL (d) CH

content in the void ratio range mentioned above. The observed resistivity results of Ca-bentonite were 11.6 and 5.4 Ohm-m at moisture contents of 10% and 40%, respectively (at 0.78 void ratio). Similar trends of variations were observed in kaolinite, CL, and CH specimens.

According to Figure 3.37, the effect of porosity was significant at low moisture content (i.e., moisture content = 10%) because of the presence of air void. However, interconnectivity of moisture with soil particles increased with the increase in moisture content and caused a substantial reduction in resistivity.

The study performed by Kibria and Hossain (2012) indicated the coupled effects of pore space and specific surface area in the electrical resistivity of soils. A scanning electron microscope (SEM) was employed to determine the microporosity of the test specimens, and then the pore spaces in the soil samples were analyzed. The SEM images of the test specimens and representative pore spaces are presented in Figure 3.38 and Figure 3.39, respectively.

The approximate percentages of pore spaces obtained from the two dimensional analyses are presented in Table 3.4.

The resistivity was plotted against pores spaces for samples 1, 2, and 4, as presented in Figure 3.40. The dry unit weights of the samples were 11.8 and 13.4 kN/m³. The moisture contents in Figure 3.40 ranged from 18% to 30%, with a 6% increment. It was observed that

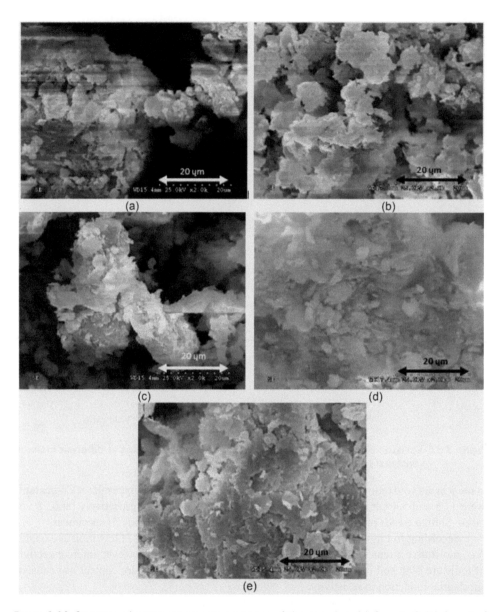

Figure 3.38 Scanning electron microscope images of the samples: (a) Sample 2 (b) Sample 3 (c) Sample 5 (d) Sample 4 (e) Sample 6 (Kibria and Hossain, 2012), with permission from ASCE

the specific surface area substantially influenced resistivity at a specific moisture content and unit weight. Test results indicated that an increase in pore percentages caused an initial increase in resistivity, followed by a reduction in resistivity. The variation was substantial at 18% moisture content and 11.8 kN/m³ dry unit weight. It was also observed that the resistivity was low at a lower specific surface area.

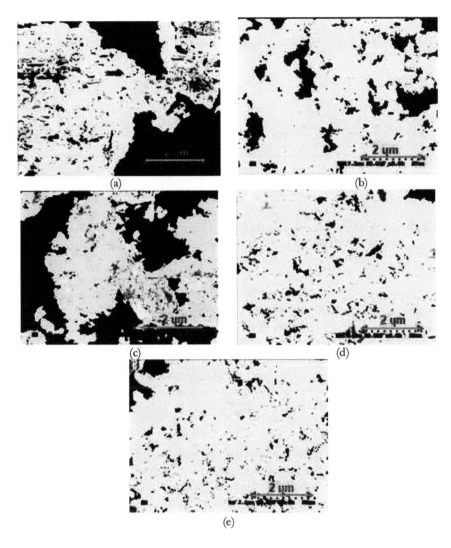

Figure 3.39 Pore space analysis of SEM images: (a) Sample 2 (b) Sample 3 (c) Sample 5 (d) Sample 4 (e) Sample 6 (Black portion is void. and white portion is soil.) (Kibria and Hossain, 2012), with permission from ASCE

Table 3.4 Percentages of pore space obtained from SEM images and specific surface area (Kibria and Hossain, 2012), with permission from ASCE

Designation of the Sample	Percentages of Pore Spaces	Specific Surface Area (m²/gm)
SAMPLE 2	27.29	75.0
SAMPLE 3	10.56	107.1
SAMPLE 5	39.01	75.0
SAMPLE 4	1.91	69.6
SAMPLE 6	2.45	82.1

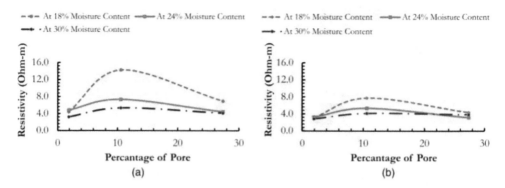

Figure 3.40 Variation of resistivity with obtained pore spaces from scanning electron images at dry unit weight: (a) 11.8 kN/m³, (b) 13.4 kN/m³, with permission from ASCE

At 18% moisture content, the resistivity might have increased because of the presence of dielectric air void and a higher specific surface area. Test results showed that the decrease in resistivity for an increase of pore space from 10.56% to 27.39% was associated with a reduction in the specific surface area. At lower (18%) moisture content, the development of a water film and moisture bridging between the particles were enhanced when the specific surface area was low, resulting in a reduction in resistivity even though the pore space increased.

3.2.13 Atterberg limits

Surface activity of a soil is related to particle size and amount of fine fraction. In addition, index properties of soils are sensitive to the specific surface area, electrolyte concentration, cation valence, and dielectric constant. Therefore, liquid limits and plasticity indices can be designated as two important indicators of physicochemical properties of soils (Mitchell and Soga, 2005).

The variations of resistivity with liquid limits at different degrees of saturation for artificial soils (Ca-bentonite and sand mixtures) obtained in the study performed by Kibria (2014) are presented in Figure 3.41. At 25% degree of saturation, resistivity decreased from 64 to 22 Ohm-m with an increase of liquid limits from 22 to 107. Similar variations were identified at 50%, 75%, and 100% degrees of saturation; however, the rates of reduction were different.

The observed variations in resistivity of the samples can be explained by the clay-water interaction phenomenon. The net negative charge of a clay structure attracts the positive area of water ions and are adsorbed. The possibility of water adsorption increases with the increase of surface charge. Therefore, moisture bridging among the particles increases with an increase in affinity to water.

The variations of resistivity with plasticity indices (PI) of the artificial samples are presented in Figure 3.42. An overall decrease in resistivity was observed for the increase of PI in each soil sample. For the increase of plasticity indices from 7 to 43, resistivity decreased as much as 1.72 times in Ca-bentonite-sand specimens at 25% degree of saturation. Resistivity reduced from 7 to 2.5 Ohm-m at 100% degree of saturation.

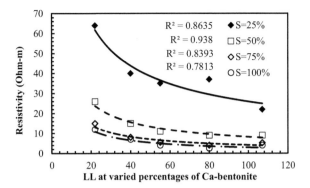

Figure 3.41 Effects of liquid limits on resistivity of artificial soils

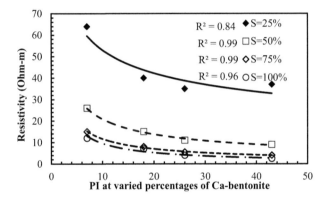

Figure 3.42 Variation of resistivity with plasticity indices of artificial soil samples (Kibria and Hossain, 2014), with permission from ASCE

3.2.14 Dielectric permittivity of soil

Electrical properties of the soil are controlled by dielectric permittivity of the soil. Dielectric permittivity is a measure of the material to store charge under applied electric field. Dielectric loss is opposite to dielectric permittivity. Dielectric loss can be defined as a measure of the proportion of the charge transferred to conduction. Saarenketo (1998) stated that the separation of electric charge can occur in four methods: electrical, molecular, orientational, and interfacial polarization. The bonding of water molecules around the soil particles is dependent upon the frequency of the current and influences the dielectric permittivity of soil. The definition of dielectric permittivity can be given by the following equation:

$$K*(\omega) = K'(\omega) - iK''(\omega) \qquad (3.11)$$

where, K' = Real part of dielectric permittivity and K" = Imaginary part of dielectric permittivity.

The author indicated that real part of the expression of dielectric permittivity might vary with natural soil constituents.

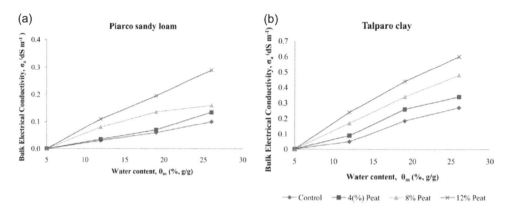

Figure 3.43 Electrical conductivity at different peat contents: (a) Picaro sandy loam (b) Talparo clay (Ekwue and Bartholomew, 2010)[10]

3.2.15 Organic content

Organic soil is usually composed of decayed materials which are intermixed with soil minerals that form a distinct texture. It retains a higher proportion of water and electrolytes, so it is highly conductive. Ekwue and Bartholomew (2010) reported on the effect of peat on the conductivity of some soils in Trinidad. They observed that an increase in the peat content increased the conductivity at constant water content and bulk density, as presented in Figure 3.43.

3.2.16 Geologic formation

Soil electrical resistivity usually exhibits a wide range of values; it is low for coastal soil and high for rocks. Studies have also demonstrated that soil resistivity is affected by geological formations. Research conducted by Giao *et al.* (2003) showed that the presence of diatom micro fossils substantially alters the geotechnical and electrical properties of clay. Robain *et al.* (1996) presented resistivity variations with the structure of pedological materials. According to the authors, low and high resistivity values are related to the macro and mesoporosity of soil.

3.3 Sensitivity of electrical resistivity with geotechnical parameters

It is evident that the electrical resistivity of soils is affected by several geotechnical parameters; however, the scale of influence differs from one parameter to another. Moisture content and unit weight significantly affect the electrical resistivity of soils. Nonetheless, the study indicates that the effect of compressibility on the electrical resistivity is in the micro-level.

---

10 Reprinted from Biosystems Engineering, Vol 108, Issue 2, E.I. Ekwue, J. Bartholomew, Electrical conductivity of some soils in Trinidad as affected by density, water and peat content,95-103, Copyright (2011), with permission from Elsevier

Based on several studies, it is apparent that increases in moisture content and unit weight cause reduction in resistivity, and vice versa. However, the study conducted by Kibria and Hossain (2012) emphasized that soil resistivity was more sensitive to moisture content than to unit weight.

Kibria (2014) conducted a study to evaluate geotechnical parameters affecting resistivity, using statistical analyses. Laboratory testing was conducted to determine the effects of soil properties on electrical resistivity. Based on the laboratory investigation of influential parameters, it was determined that moisture content; dry unit weight; void ratio; degree of saturation; and clay properties such as cation exchange capacity, specific surface area, and activity significantly affect the electrical resistivity of soil. Pore water properties, mineral composition, and ion composition also influence the electrical resistivity. Kibria (2014) conducted statistical analyses on the results of the laboratory tests and observed that the statistical significances were different for various parameters when correlated with resistivity.

Based on the available studies, a sensitivity matrix can be developed, as presented in Table 3.5. It should be noted that the sensitivity matrix presented herein provides an indication of the importance of the affecting parameters and should be considered approximate. Since resistivity is a complex phenomenon and depends on many factors, the level of sensitivity is subject to change, depending on the soil conditions.

Any soil resistivity model should consider these parameters. However, some of these parameters are correlated, and using all of the parameters in a single model can cause multicollinearity in statistical analyses if statistical methods are employed to develop the model. Additionally, ionic compositions, fabric study, mineralogical study, and pore water composition tests are not typically performed in geotechnical investigations.

A practically applicable model should be statically verified, simple, and able to evaluate physical properties. Chapter 4 presents the available electrical mixing models, their

Table 3.5 Sensitivity of parameters on electrical resistivity of soils

Soil Parameters	Level of Sensitivity – Primary	Level of Sensitivity – Secondary
Moisture Content	√	–
Unit Weight	√	–
Degree of Saturation	√	–
Volumetric Moisture Content	√	–
Compaction Condition	√	–
Void Ratio	√	–
Minerology	√	–
Ion Composition	–	√
Structure, Packing and Hydraulic Conductivity	–	√
Cation Exchange Capacity (CEC)	√	–
Specific Surface Area (SSA)	√	–
Temperature	√	–
Atterberg Limits	–	√
Compressibility	–	√
Dielectric Permittivity of Soil	–	√
Organic Content	–	√
Geologic Formation	–	√
Soil Fabric	–	√

applicability, and the requirements of a practically applicable model to bridge the gap that currently exists between geotechnical and geophysical engineering. Additionally, Chapter 4 describes details of practically applicable models developed by Kibria and Hossain (2015 and 2016) and presents the correlations of electrical resistivity with various geotechnical properties developed from the models.

Electrical mixing models

Bridging the gap between geophysical and geotechnical engineering

4.1 General

In 1942, G. E. Archie presented a model which correlated electrical conductivity of soil with porosity and pore fluid conductivity. The objective of his model was to aid exploration of oil and gas reservoirs. Since 1942, significant studies have been conducted to develop electrical mixing models for various earth materials. At present, many electrical mixing models are available to explain the relationship between electrical resistivity and soil properties. This chapter presents a review of the available electrical mixing models, their applicability and limitations, descriptions of practically applicable models developed by Kibria and Hossain (2015 and 2016), and an evaluation of geotechnical parameters, using Kibria and Hossain's (2015 and 2016) models.

4.2 Available electrical mixing models

Archie (1942) developed an empirical formula to correlate the bulk resistivity of saturated soil with the pore fluid resistivity and porosity. If the resistivity of soil is ρ, resistivity of pore fluid is ρ_w and porosity is n, then Archie's formula can be stated as:

$$\rho = a.\rho_w.n^{-m} \tag{4.1}$$

where, a and m are the fitting and cementation parameters, respectively. According to the study, the value of m depends on the interconnectivity of the pore network and tortuosity. In an unsaturated media, Archie's law can be presented as:

$$\rho = a.\rho_w.n^{-m}.S^{-B} \tag{4.2}$$

where, S is the degree of saturation.

Archie's law was developed based on the sandy soil; therefore, the role of surface charge in electrical conduction of clayey soil was not described in the model. However, surface charge has been reported as an important parameter in the electrical conduction of fine-grained soil. Therefore, several theoretical and experimental investigations have been performed to explain the electrical conduction in clayey soil.

Sauer et al. (1955) suggested flow of current through soils, using a three-element network model. According to the authors, current flows through the surface charge and pore fluid along a combined series and parallel paths in clayey soils. The conductivity of a clayey particulate media (σ) can be described by the following equation:

$$\sigma = \frac{a.\sigma_s.\sigma_W}{(1-e).\sigma_W + e.\sigma_s} + b.\sigma_s + c.\sigma_W \tag{4.3}$$

where σ_s is the surface conductance, σ_w is the pore water conductivity, constants a, b, c, d, and e are functions of porosity and degree of saturation. The current flow paths and equivalent circuit diagram of the three-element network model is illustrated in Figure 4.1. Waxman and Smits (1968) developed a simplified model, where the series effects of surface conductance and pore fluid conductivity were not considered. The contribution of series path was substituted by a constant surface conductance. The system was equivalent to a circuit composed of two resistors connected in a parallel direction; therefore, the proposed expression is known as two-parallel resistor model. According to the model, the electric conductivity of a soil (σ) can be expressed as:

$$\sigma = X.(\sigma_S + \sigma_W) \tag{4.4}$$

where, X is a constant that is reciprocal to formation factor.

Shah and Singh (2005) suggested a generalized form of Archie's law for fine-grained soil. According to the authors, the effects of surface conductivity can be included in the cementation factor of Archie's law. Hence, the proposed relationship in terms of conductivity can be expressed as:

$$\sigma = c.\sigma_w.\theta^n \tag{4.5}$$

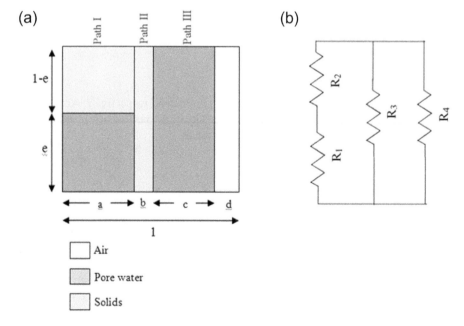

Figure 4.1 Three-element network model: (a) possible current flow paths and (b) equivalent resistance

where c = 1.45 when CL<5% and c =0.6 CL$^{0.55}$ when CL≥5%, m = 1.25 when CL<5% and m = 0.92 CL$^{0.2}$ when CL≥5%, σ is bulk conductivity of soil, σ$_w$ is pore water conductivity, θ is volumetric moisture content, and CL is percentage of clay fraction in soil.

In many geophysical applications, the relationship between bulk soil resistivity and pore water resistivity is measured by using a formation factor (F). A formation factor can be defined by the ratio of bulk resistivity (ρ) and pore water resistivity (ρ$_w$), in accordance with the following equation:

$$F = \frac{\rho}{\rho_w}$$ (4.6)

It is evident from the electrical mixing models that clay soil is more conductive than sandy soil. However, saturated sandy soil may demonstrate lower resistivity than dry compacted clay. Because of these factors, overlapping of resistivity values is observed for different types of soils.

4.3 Applicability and limitations of the available models

The available electrical mixing models of clayey soil describe resistivity as a function of pore fluid conductivity, pore connectivity, and surface conductance. The measurement of pore water conductivity requires extraction of moisture from the soil, and such methods are difficult to perform in a regular investigation. Furthermore, determination of surface conductance is time intensive. The pore connectivity is defined by tortuosity, which correlates with the conductive path and length of soil sample (Mitchell and Soga, 2005). An accurate estimation of tortuosity requires fundamental information about the packing of soil samples. Due to the inherent complexities of the measurement methods of the various parameters of the existing electrical mixing models, these models are not typically used in the engineering practices. A practically applicable model is required to overcome the current limitations associated with the electrical mixing models and soil parameters estimated from electrical resistivity.

4.4 Practically applicable models (Kibria and Hossain, 2015 and 2016)

Kibria and Hossain (2015 and 2016) developed electrical resistivity models to correlate resistivity with influential geotechnical properties of compacted and undisturbed clay specimens, using multiple linear regression (MLR) analyses. An experimental program was developed to investigate the relationship between electrical resistivity and soil properties. The test specimens included: (1) undisturbed soil, (2) disturbed soil, (3) clay minerals, and (4) artificial soils. The natural soil specimens included fat clay (CH) and lean clay (CL), and clay minerals consisted of Ca-bentonite and kaolinite. The artificial soil samples were prepared using Ca-bentonite and fine sands at different weight percentages.

The engineering properties of the soil specimens were determined using conventional geotechnical tests and advanced methods, such as a scanning electron microscope, energy dispersion spectroscopy, and ion chromatography. After the soil characterization, the electrical resistivity tests were conducted on compacted and undisturbed soil samples at varying geotechnical conditions.

As noted previously, Kibria and Hossain (2015 and 2016) used the MLR method to develop the models. According to Kutner et al. (2005), the predictor variables of a MLR model should not be correlated; they should be free of multicollinearity. However, practically, the predictor variables are often correlated with each other. The basic interpretation of regression parameters, i.e., changes in expected results for unit changes in a predictor variable, may not be appropriate when predictor variables are correlated with each other.

Soil resistivity results showed that the gravimetric moisture, dry unit weight, and void ratio simultaneously affect the electrical resistivity of soils. However, to avoid potential multicollinearity, Kibria and Hossain (2015 and 2016) did not consider all of these parameters in their model development. The weight-volume relationship of soil indicates direct relationships among moisture content, dry unit weight, and void ratio. Thus, Kibria and Hossain (2015 and 2016) considered gravimetric moisture content, void ratio, and dry unit weight, using degrees of saturation to avoid multicollinearity in the models.

To include the effect of clay properties in the models, Kibria and Hossain (2015 and 2016) evaluated parameters such as cation exchange capacity (CEC), specific surface area, activity, Atterberg limits, etc. According to Shah and Singh's study (2005), the clay fraction can be linearly correlated with the CEC of the soils. Shewartz *et al.* (2008) conducted a study to quantify field-scale moisture content using electrical resistivity imaging. To avoid the potential difficulty of moisture extraction from the soil, exchangeable cations (Ca/Mg) were used, as a proxy of pore water properties, to calibrate Archie's Law in site-specific conditions. Additionally, Kibria (2014) indicated that percentages of clay mineral and pore water properties were linearly correlated with CEC. Therefore, Kibria and Hossain (2015 and 2016) considered CEC in their models to represent the effect of clay properties in electrical conduction of soils.

Since the objective of the Kibria and Hossain (2015 and 2016) models was to correlate resistivity with physical properties of soils, the ionic compositions and pore water ions were not used in the model development.

4.4.1 Compacted clay model (Kibria and Hossain, 2015)

The electrical resistivity model developed by Kibria and Hossain (2015) for compacted clay can be presented as:

$$R^{-0.25} = 0.43398 + 0.00309 S_R - 14.35204 \text{CEC}^{-1.5} \tag{4.7}$$

where, R = Electrical resistivity (Ohm-m) corrected at 15.5 deg. temperature according to ASTM G187, S_R = Degree of saturation (%), CEC = Cation exchange capacity (cmol+/kg). Range of the model: R = [2.6, 504.3], S_R = [21.8, 100], CEC = [13.3, 79.0].

It should be noted that the coefficient of regression of the model was 90.1%. Therefore, 90.1% of the variation in resistivity (Ohm-m) of compacted clays was explained by the degree of saturation and CEC for the dataset used in the model development.

4.4.1.1 Laboratory validation of compacted clay model (Kibria and Hossain, 2015)

The experimental results of artificial soil samples were utilized to compare the model's predicted results. A total of four artificial soils were used for the model validation. Electrical resistivity tests were conducted at varying degrees of saturation, and the observed results were corrected for 15.5°C temperature. A comparison of the model's predicted and observed resistivity is presented in Figure 4.2. It is evident that the model's performance

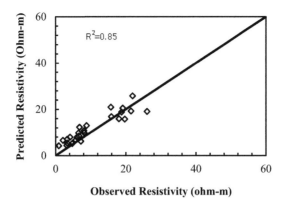

Figure 4.2 Comparison of model-predicted-and-observed resistivity (Kibria and Hossain, 2015)[1]

was satisfactory in the laboratory environment; however, it should also be satisfactory in the field to be recognized as a practically applicable model. Kibria and Hossain (2015) used this model to estimate geotechnical parameters during construction of a compacted clay liner.

4.4.1.2 Field validation of compacted clay model (Kibria and Hossain, 2015)

Three RI tests were performed parallel to each other and were combined to develop a quasi-3D profile. Thereafter, a horizontal resistivity contour was determined at clay liner depth, using the 3D resistivity profile. Based on the observed resistivity and CEC of *in-situ* soils at different locations of the clay liner, degrees of saturation were calculated from the model. The model-predicted results were compared with the measured-in-place degrees of saturation.

4.4.1.2.1 DESCRIPTION OF THE STUDY AREA

The City of Denton municipal solid waste (MSW) landfill is located 40 miles northwest of Dallas, Texas. The conventional operation of the landfill began in 1984; however, the facility started functioning as an enhanced leachate recirculated (ELR) system in May 2009. The City of Denton Landfill has been subjected to expansion since 1984 (Manzur, 2013). The footprint of the landfill is divided into different cells, as presented in Figure 4.3. During the study (September, 2013), Cell 4A was under construction.

It should be mentioned that the construction of the clay liner was not completed during the RI tests on 16 September 2013. At the foundation soil, 45-cm clay liners were constructed in three lifts, and 20-cm loosely compacted clayey soils were placed on top of the clay liner. The top soil layer provided a working platform for the conduction of the RI tests and ensured further resistance to potential leakages in the liner system. Additional layers were proposed to be placed on top of the loosely compacted soils. The cross section of the liner during the RI tests is presented in Figure 4.4.

1 Reprinted from Waste Management, Vol. 39, Golam Kibria, Md. Sahadat Hossain, Investigation of degree of saturation in landfill liners using electrical resistivity imaging, 197-204, Copyright (2015), with permission from Elsevier

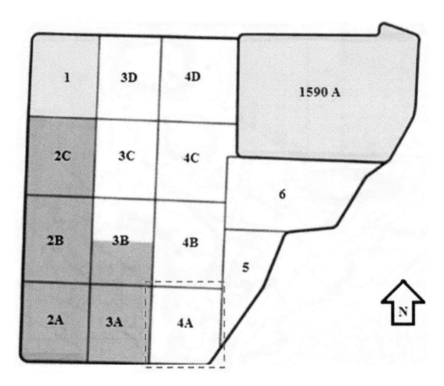

Figure 4.3 Cells of City of Denton MSW landfill (Kibria and Hossain, 2015)[2]

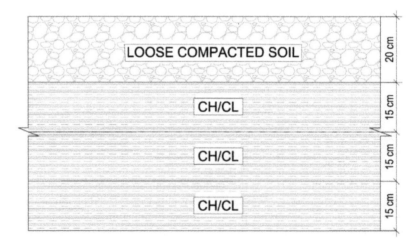

Figure 4.4 Cross section of the clay liner during RI tests (as of 16 September 2013)

2 Reprinted from Waste Management, Vol. 39, Golam Kibria, Md. Sahadat Hossain, Investigation of degree of saturation in landfill liners using electrical resistivity imaging, 197-204, Copyright (2015), with permission from Elsevier

Figure 4.5 Super Sting R8/IP multichannel system and switch box (AGI, Austin, Texas)

RI tests were conducted using the Super Sting R8/IP multichannel system manufactured by Advanced Geosciences Inc. (AGI), Austin, Texas. The equipment measures apparent resistivity up to 8 points for a single current injection. Four passive cables, each containing 14 takeouts, were attached with the resistivity meter. The power supply of the equipment was provided by a 12V battery. A switch box was utilized to develop a close form circuit with the electrodes, cable, and Super Sting R8/IP equipment. The spacing between electrodes can be as much as 6 m; however, an increase in spacing may result in poor image resolution. The investigative depth was approximately 20% of length of the profile. Although the RI tests can be performed utilizing 56 electrodes, the equipment allows using 28 electrodes when site access is limited and the required area of investigation is minimal. The RI equipment and switch box are illustrated in Figure 4.5.

4.4.1.2.2 RI TESTS IN THE COMPACTED CLAY LINER

2D resistivity imaging (RI) tests were conducted on Cell 4A after the completion of three lifts and placement of 20 cm of loosely compacted soil. A total of 28 electrodes with a spacing of 1.52 m were used in the RI tests. The length of each profile was 41.2 m, as presented in Figure 4.6. The distances of profile CL1 from the end of Cell 3 and the access road were 45.6 and 30.5 m, respectively. The RI profile of CL2 and CL3 were located 22.9 and 53.4 m apart from Line CL1. Since dipole-dipole has the advantages of low electromagnetic coupling and high sensitivity in response to the variation in horizontal direction (Loke, 2000), this method was utilized to conduct RI tests.

Figure 4.6 Location of RI tests: (a) Cell 4A (photograph taken from top of Cell 3) and (b) schematic of RI profiles (Kibria and Hossain, 2015)[3]

Figure 4.7 Operational setup of RI test in compacted clay liner: (a) location of tests (b) RI equipment and switch box (c) layout of electrodes and cables (d) sealing of insertion point ((Kibria and Hossain, 2015)[4]

3 Reprinted from Waste Management, Vol. 39, Golam Kibria, Md. Sahadat Hossain, Investigation of degree of saturation in landfill liners using electrical resistivity imaging, 197-204, Copyright (2015), with permission from Elsevier.
4 Reprinted from Waste Management, Vol. 39, Golam Kibria, Md. Sahadat Hossain, Investigation of degree of saturation in landfill liners using electrical resistivity imaging, 197-204, Copyright (2015), with permission from Elsevier.

The electrode penetration depth was less than 20 cm during RI tests. Once the RI tests were completed, the insertion points of the electrodes were sealed with bentonite. The operational setup of the RI tests in the clay liners is presented in Figure 4.7.

4.4.1.2.3 2D RESISTIVITY IMAGING RESULTS

EarthImager 2D software was utilized for the data analysis and resistivity image construction. At the initial condition, a smooth inversion method was adopted for the analysis of measured apparent resistivity. The finite element method, with Cholesky decomposition equation solver, was employed for forward modeling. The Dirichlet boundary condition was used in the forward model. The RMS error of 3% was considered for the stopping criteria of the iteration.

The RI images along CL1, CL2, and CL3 are shown in Figure 4.8. At surface, the resistivity was as much as 15 Ohm-m. However, resistivity decreased substantially between depths of 0.5 to 1.2 m, and the resistivity was below 6.5 Ohm-m up to a depth of 2.5 m. The observed resistivity was over 10 Ohm-m below 2.5 m of the surface.

The observed variations in resistivity can be explained by the existing subsurface conditions. At the surface, resistivity was high because of the presence of loosely compacted soil. The resistivity decreased between depths of 0.5 and 1.2 m because of the presence of compacted clay lifts. According to Qian *et al.* (2002), clay liners should be constructed at a minimum of 95% of the maximum dry unit weight at Standard Proctor, and 0% to 4% wet-of-optimum moisture content. Therefore, the observed resistivity was low because of the high

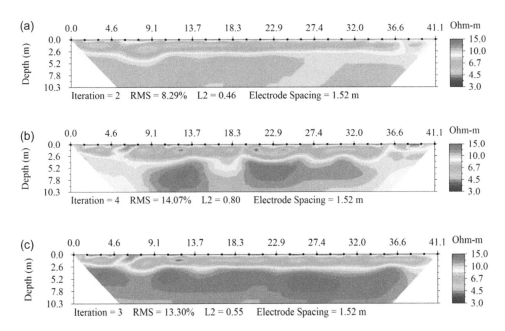

Figure 4.8 2D resistivity image on compacted clay liner along (a) CL1, (b) CL2, (c) and CL3 (Tests were conducted on 16 September 2013) (Kibria and Hossain, 2015)[5]

5 Reprinted from Waste Management, Vol. 39, Golam Kibria, Md. Sahadat Hossain, Investigation of degree of saturation in landfill liners using electrical resistivity imaging, 197-204, Copyright (2015), with permission from Elsevier.

degree of saturation in this zone. Another important observation was the effect of compaction below the clay liners. The voids in the subgrade soil might have decreased due to the construction of the lifts, causing a reduction in resistivity. Therefore, the influence of compaction beneath the clay liner was also identified by using the RI test. According to the investigation, the effect of compaction extended up to 2.5 m.

4.4.1.2.4 DETERMINATION OF QUASI-3D AND HORIZONTAL RI PROFILE AT CLAY LINER

A quasi-3D RI section was developed, using the parallel 2D profiles. In a quasi-3D section, the measured 2D resistivity values were merged into a 3D data format. Then the new 3D data file was analyzed using EarthImager 3D software to obtain a quasi-3D image. During the analyses, the forward and inversion modeling parameters were similar to the 2D data analyses; however, the stabilizing and damping factors were set to high values (equal to 1000) to minimize model roughness. The observed 3D resistivity profile is presented in Figure 4.9.

The 3D resistivity profile was utilized to determine the horizontal resistivity contour at the depth of the clay liner. The horizontal slice was developed, using the static slice option of EarthImager 2D software. The horizontal resistivity profile at 0.31 m depth below the surface is presented in Figure 4.10. It was observed that the resistivity was in between 5 to 8.5 Ohm-m at this depth. To quantify electrical resistivity at different locations, the horizontal profile was divided into several grids, as presented in 4.10 (b). At each grid point, resistivity was measured from the profile.

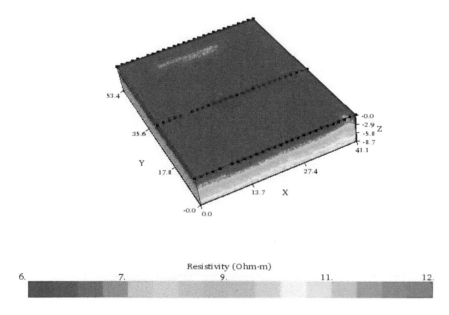

Figure 4.9 Quasi-3D resistivity profile of the investigated area (Kibria and Hossain, 2015)[6]

6 Reprinted from Waste Management, Vol. 39, Golam Kibria, Md. Sahadat Hossain, Investigation of degree of saturation in landfill liners using electrical resistivity imaging, 197-204, Copyright (2015), with permission from Elsevier.

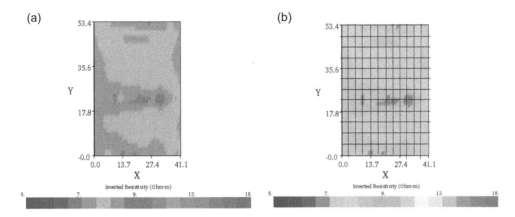

Figure 4.10 Horizontal resistivity profile at depth of clay liner: (a) at 0.31 m /1 ft from surface and (b) dividing the profile into grids (Kibria and Hossain, 2015)[7]

4.4.1.2.5 COMPARISON OF MODEL PREDICTED DEGREE OF SATURATION

As a part of the construction quality control, *in-situ* tests were performed on the compacted clay liners. The coordinates of test locations were plotted in Cell 4A of the Google map, as illustrated in Figure 4.11. Figure 4.11 indicates that the *in-situ* tests were carried out in several locations of the cell, but the test results under the boundary of 3D RI profile were utilized for the comparison.

The coordinates of the horizontal profile and the *in-situ* tests were matched to identify the resistivity at the locations of in-place density tests. The Atterberg limits and Standard Proctor compaction tests were also carried out on the collected soil samples from the test locations. A summary of the soil properties utilized in the comparison is presented in Table 4.1.

To evaluate the model performance, the observed liquid limits were utilized to calculate CEC (cmol+/kg) of the samples, using the correlation of Yukselen and Kaya (2006):

$$CEC = 0.2027LL + 16.231 \tag{4.8}$$

As the horizontal resistivity profile was located at 0.31 m beneath the surface, the temperature at 0.31 m depth was assumed to be similar to the ambient temperature (16.2 deg. C). Therefore, quantified resistivity values were corrected with respect to 16.2°C according to ASTM G187. The calculated CEC and corrected resistivity values were used to predict degrees of saturation, using the developed model.

The predicted degrees of saturation were compared with the measured values at the site condition. The in-place density and moisture content were incorporated with specific gravity (2.65 for all cases) to obtain degrees of saturation at test locations. Thereafter, model-predicted results were compared with the measured degrees of saturation in the field. The

7 Reprinted from Waste Management, Vol. 39, Golam Kibria, Md. Sahadat Hossain, Investigation of degree of saturation in landfill liners using electrical resistivity imaging, 197-204, Copyright (2015), with permission from Elsevier.

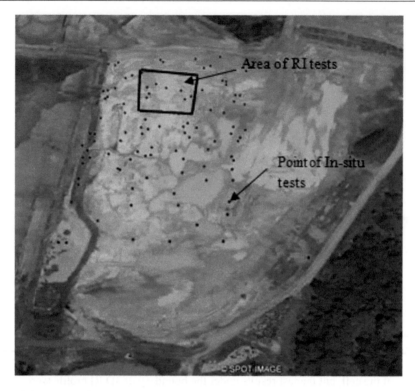

Figure 4.11 Location of *in-situ* tests for compaction quality control (@Google Maps, 2013)

Table 4.1 Soil properties at locations of *in-situ* tests (data provided by City of Denton Landfill authority, 16 September 2013 (Kibria and Hossain, 2015)[8]

Test	Coordinates		In-situ test		LL	PI	Soil Type	CEC (cmol+/kg)	Relative density
	N	W	Moisture content (%)	Dry Unit wt. (kN/ m³)					
33	33.1864	−97.081	15.7	17.8	34	22	CL	23.12	97.10%
125	33.1862	−97.081	17.5	17.4	49	33	CL	26.16	99.90%
135	33.1861	−97.081	18.1	16.8	52	36	CH	26.77	99.90%
42	33.1862	−97.081	18.3	16.9	37	23	CL	23.73	99.30%
36	33.1863	−97.081	15.5	18.3	34	22	CL	23.12	99.70%
66	33.1862	−97.08	17.2	17.7	34	22	CL	23.12	96.50%
127	33.1861	−97.08	19.7	17	49	33	CL	26.16	97.90%
41	33.1861	−97.08	19.4	16	52	36	CH	26.77	95.10%

8 Reprinted from Waste Management, Vol. 39, Golam Kibria, Md. Sahadat Hossain, Investigation of degree of saturation in landfill liners using electrical resistivity imaging, 197-204, Copyright (2015), with permission from Elsevier.

Table 4.2 Summary of comparison (Kibria and Hossain, 2015)[9]

In-situ Test ID	Observed Degree of Saturation (%)	Predicted Degree of Saturation (%)	Variation (%)
33	90.8	88.7	2.1
125	93.2	90.5	2.7
135	88	78.4	9.6
42	89.4	82.2	7.2
36	98.3	97.5	0.8
66	97.9	97.5	0.4
127	98.9	90.5	8.4
41	82.6	87.4	−4.8

summary of the comparison is presented in Table 4.2. Table 4.2 indicates that the prediction error ranged from 0.4% to 9.7%; therefore, the performance of the model was quite satisfactory for the current dataset.

4.4.2 Evaluation of compacted clay properties using Kibria and Hossain's (2015) model

Based on the model validation, it is evident that the model was able to predict geotechnical parameters from RI. A set of geotechnical parameters were evaluated from the Kibria and Hossain (2015) compacted clay model and are presented in the following subsections. During evaluation of the geotechnical properties, cation exchange capacity (CEC) was empirically estimated, using the correlation proposed by Yukselen and Kaya (2006). It should be noted that the geotechnical properties are only valid for the compacted clay model assumptions and within the parameter ranges applicable for that model. Additionally, the field resistivity should be corrected for 15.5°C temperature according to ASTM G187 prior to using correlation charts to determine geotechnical parameters.

4.4.2.1 Evaluation of moisture content

Based on the phase relationship of soils, the degree of saturation is a function of moisture content. Moisture content can be determined for a given dry unit weight and specific gravity when the degree of saturation is known. The following equation can be employed to evaluate moisture content from Kibria and Hossain's (2015) compacted clay model:

$$w = \frac{\left\{R^{-0.25} + 14.35204(0.2027LL + 16.231)^{-1.5} - 0.43398\right\}}{0.00309} X\left(\frac{Gs\Upsilon w}{\Upsilon d} - 1\right)\frac{1}{Gs} \quad (4.9)$$

The following example presents the analytical approach to determining moisture content from resistivity of compacted clay soil. If the observed resistivity from a RI test in a compacted

9 Reprinted from Waste Management, Vol. 39, Golam Kibria, Md. Sahadat Hossain, Investigation of degree of saturation in landfill liners using electrical resistivity imaging, 197-204, Copyright (2015), with permission from Elsevier.

fat clay (CH) soil is 10.0 Ohm-m at 1.5 m depth, the liquid limit (LL) of the soil is 75, the approximate dry unit weight is 16.5 kN/m³, and the specific gravity is 2.65, the estimated moisture content of the soil is:

$$w = \frac{\left\{10^{-0.25} + 14.35204\left(0.2027x75 + 16.231\right)^{-1.5} - 0.43398\right\}}{0.00309} X \left(\frac{2.65x9.81}{16.5} - 1\right)\frac{1}{2.65}$$

$$w = 14.75\%$$

The moisture content obtained from the equation is in good agreement with the chart in Figure A1-A4 in Appendix A and is applicable for this sample.

4.4.2.2 Evaluation of dry unit weight

Since the dry unit weight of soil is correlated with the degree of saturation, this parameter can be evaluated from the resistivity of soil, using Kibria and Hossain's (2015) model. At a given degree of saturation, specific gravity, moisture content, and dry unit weight can be determined analytically from resistivity by using the Kibria and Hossain (2015) model in accordance with the following equation:

$$\left(\frac{1}{\Upsilon d}\right) = \left[\frac{0.00309.w.Gs}{\left\{R^{-0.25} + 14.35204\left(0.2027LL + 16.231\right)^{-1.5} - 0.43398\right\}} + 1\right]\left(\frac{1}{Gs\Upsilon w}\right) \quad (4.10)$$

The following example shows how to determine the unit weight of compacted clay soils, using resistivity. If the observed resistivity from a RI test in a typical fat clay (CH) soil is 10.0 Ohm-m at 1.5 m depth, the LL of the soil is 75, the observed moisture content from the geotechnical investigation is 20%, and the specific gravity is 2.65, the estimated unit weight of the sample is:

$$\left(\frac{1}{\Upsilon d}\right) = \left[\frac{0.00309x20x2.65}{\left\{10^{-0.25} + 14.35204\left(0.2027x75 + 16.231\right)^{-1.5} - 0.43398\right\}} + 1\right]\left(\frac{1}{9.81x2.65}\right)$$

$$\Upsilon d = 14.6kN / m^3$$

According to the presented equation, the unit weight of the sample is 14.6 kN/m³. The estimated unit weight from the equation is in agreement with the provided correlation chart presented in Figure A5 to Figure A7 in Appendix A, which is applicable for this sample.

4.4.2.3 Evaluation of degree of saturation

Kibria and Hossain's (2015) compacted clay model suggests that the degree of saturation can be evaluated from the resistivity of soil. If the LL of a clayey soil is estimated from CEC, then resistivity is a function of the degree of saturation. Hence the Kibria and Hossain (2015) model can be used to determine the degree of saturation, using the following equation:

$$S_R = \frac{R^{-0.25} - 0.43398 + 14.35204x\{0.2027LL + 16.231\}^{-1.5}}{0.00309} \quad (4.11)$$

The estimation of the degree of saturation using resistivity is presented in the following equation. If the observed resistivity from a RI test of a compacted fat clay (CH) soil is 10.0 Ohm-m at 1.5 m depth and the LL and specific gravity of the soils are 75 and 2.65, respectively, the degree of saturation of the sample is:

$$S_R = \frac{10^{-0.25} - 0.43398 + 14.35204x\{0.2027x75 + 16.231)\}^{-1.5}}{0.00309}$$

$$S_R = 67.9\%$$

The presented equation indicates that the degree of saturation of the sample is 67.9%; therefore, the estimated degree of saturation is in agreement with the correlation chart presented in Figure A8 in Appendix A.

4.4.2.4 Evaluation of void ratio

Degree of saturation is a function of void ratio according to the phase relation of soils, and the void ratio can be determined for a given degree of saturation and specific gravity. An analytical approach can be used to determine the void ratio by using resistivity and Kibria and Hossain's (2015) compacted clay model. If the degree of saturation parameter is presented as a function of the void ratio, the following equation can be developed which correlates void ratio with resistivity:

$$e = \frac{0.00309.wGs}{R^{-0.25} - 0.43398 + 14.35204(0.2027LL + 16.231)^{-1.5}} \tag{4.12}$$

An example is presented to show how to estimate void ratio of soils using resistivity. If the observed resistivity from a RI test in a compacted fat clay (CH) soil is 10.0 Ohm-m at 1.5 m depth, the liquid limit (LL) of the soil is 75, the observed moisture content from the geotechnical investigation is 20%, and the specific gravity is 2.65, the void ratio of the sample is:

$$e = \frac{0.00309x20x2.65}{10^{-0.25} - 0.43398 + 14.35204(0.2027x75 + 16.231)^{-1.5}}$$

$$e = 0.78$$

The equation indicates that the void ratio of the sample is 0.78; therefore, the estimated void ratio is in agreement with the correlation chart presented in Figure A9 to Figure A11 in Appendix A, and it is applicable for this sample.

4.4.2.5 Evaluation of compaction level

Compaction level is one of the most important parameters in geotechnical and geoenvironmental construction because the performance of the structures is related to the compaction level. Geotechnical parameters such as swelling potential, future settlement, bearing capacity, and permeability dictate the performance of a structure and are dependent on the compaction condition. The compaction level can be evaluated by the equation presented below if the dry unit weight is determined, using resistivity, and the maximum dry density and optimum moisture content of the soil sample are known.

$$\left(\frac{1}{\Upsilon d}\right)=\left[\frac{0.00309.w(optimum).Gs}{\left\{R^{-0.25}+14.35204\left(0.2027LL+16.231\right)^{-1.5}-0.43398\right\}}+1\right].\left(\frac{1}{Gs\Upsilon w}\right) \qquad (4.13)$$

$$Compaction\,Level=\left(\frac{\Upsilon d}{\Upsilon max}\right) \qquad (4.14)$$

The following example shows how to estimate the compaction level of soils, using resistivity. If the observed resistivity from a RI test in a typical lean clay (CL) soil is 10.0 Ohm-m at 1.5 m depth, the liquid limit (LL) of the soil is 40, and the assumed optimum moisture content (OMC) and maximum dry density of the sample are 18% and 16.5 kN/m³, respectively, the compaction level of the sample is:

$$\left(\frac{1}{\Upsilon d}\right)=\left[\frac{0.00309x18x2.65}{\left\{10^{-0.25}+14.35204\left(0.2027x40+16.231\right)^{-1.5}-0.43398\right\}}+1\right].\left(\frac{1}{2.65x9.81}\right)$$

$$\Upsilon d=16.3kn/m3$$

$$Compaction\,Level=\left(\frac{16.3}{16.5}\right)=0.99$$

The observed compaction level is 0.99; therefore, the obtained compaction level is in agreement with the correlation chart presented in Figure A12 to Figure A15 in Appendix A, applicable for this sample.

4.4.2.6 Evaluation of cation exchange capacity

The Kibria and Hossain (2015) compacted clay model emphasized that resistivity is a function of CEC at a specific degree of saturation. It is an important parameter for soil scientists to use to determine fertility and ionic activity, which are related to plant nutrition. Subsoil contamination can be indirectly evaluated from CEC since the ionic activity is subjected to change in the presence of contaminated ions. It should be noted that the CEC should be evaluated at fully saturated condition since the resistivity is highly influenced by ionic properties at saturated condition. This parameter can be evaluated, using resistivity, as follows:

$$CEC^{-1.5}=\frac{R^{-0.25}-0.43398-0.00309S_R}{-14.35204} \qquad (4.15)$$

At fully saturated condition, S_R = 100%, and the equation becomes:

$$CEC^{-1.5}=\frac{R^{-0.25}-0.43398-0.309}{-14.35204} \qquad (4.16)$$

If the resistivity of a clayey soil is 6.0 Ohm-m at fully saturated condition, then the CEC of the sample is:

$$CEC^{-1.5} = \frac{6^{-0.25} - 0.43398 - 0.309}{-14.35204}$$

$$CEC = 26.7 cmol + /kg$$

The observed compaction level is 26.7 cmol+/kg; therefore, the obtained compaction level is in agreement with the design chart in Figure A16 in Appendix A, applicable for this sample.

4.4.3 Undisturbed clay model (Kibria and Hossain, 2016)

Electrical resistivity models for undisturbed soils were developed using an approach similar to that described for compacted clays. Two models were developed for undisturbed soil samples, as presented below.
  Two-parameter model:

$$R^{-0.75} = -0.13063 + 0.00304 S_R + 0.00387 CEC \tag{4.17}$$

One-parameter model:

$$R^{-0.75} = -0.0106 + 0.00299 S_R \tag{4.18}$$

Here, Y = Electrical resistivity (Ohm-m) of undisturbed soil samples corrected at 15.5°C temperature in accordance with ASTM G187, S_R =Degree of saturation (%), CEC = Cation exchange capacity (CEC) (cmol+/kg). Range of the model: R = [4.2, 49.4], S_R= [29.1, 100], CEC = [23.9, 40.1].
  The two-parameter model explained 83.6% of the variations in resistivity with the degree of saturation and CEC input parameters, The coefficient of regression was 77.3% in the model, correlating resistivity only with the degree of saturation (one-parameter model).

4.4.3.1 Laboratory validation of the undisturbed model (Kibria and Hossain, 2016)

The experimental results of a high plasticity clay specimen were utilized to validate the resistivity models of undisturbed soil specimens. The CEC of the sample was 40.9 cmol+/kg, and degrees of saturation varied from 51% to 94% during resistivity tests on this specimen. Both one-and-two-parameter models were used to estimate electrical resistivity, as presented in Figure 4.12. The maximum errors in estimates were 4.4 and 1.7 Ohm-m in the one and two-parameter models, respectively. It is evident that the models' performances were satisfactory in the laboratory scale; however, they should be satisfactory in the field scale also to be recognized as practically applicable models. Kibria and Hossain (2016) used these models to estimate degree of saturations in slopes.

4.4.3.2 Field validation of undisturbed clay model (Kibria and Hossain, 2016)

The performance of MLR models correlating resistivity of undisturbed soil samples with degree of saturation and CEC was evaluated, using RI tests, in the slopes along highway Loop 12 and US 287 south in Dallas, Texas. The RI test was performed the day before a soil

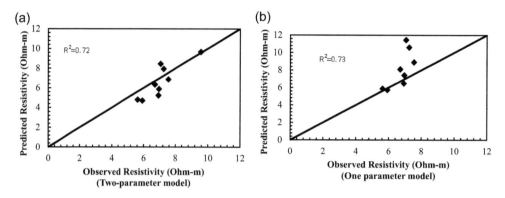

Figure 4.12 Comparison of estimated resistivity with experimental observations: (a) Two-parameter model and (b) One-parameter model (Kibria and Hossain, 2016), with permission from ASCE

test boring in the slope along highway Loop 12. The observed electrical resistivity results along the boreholes were utilized to predict the degree of saturation, using MLR models. Moisture sensors were installed for the determination of the active zone in the US 287 slope (Hossain, 2012). The RI tests were conducted along the moisture sensors, and the observed data was used for the comparison of predicted degrees of saturation with field results.

4.4.3.2.1 RI AT LOOP 12

2D multi-electrode arrays were utilized for RI tests, using Super Sting R8/IP equipment. The spacing of the electrodes was selected based on the required resolutions, sizes of objects under investigation, and depth of penetration required for the site investigations. RI was performed at the crest of the slope along highway Loop 12, using dipole-dipole array. A total of 56 electrodes were placed at a spacing of 1.52 m; therefore, a 84 m long profile was considered in the investigation. EarthImager 2D software was utilized for the modeling and image construction. Based on an initial smooth inversion model, forward and inversion modeling were conducted on the measured apparent resistivity. The finite element model adopted for forward modeling consisted of the Cholesky decomposition equation solver and Dirichlet boundary condition. The RMS error of 3% was considered for the stopping criteria of the iteration. RI at the crest of the slope is presented in Figure 4.13.

A low resistivity zone was at depths between 2.15 to 7 m, as indicated by the parallel dotted lines in Figure 4.13. The resistivity in this zone was approximately less than 10 Ohm-m. However, a substantial increase in resistivity was identified at a depth of 8.5 m below the ground surface. The obtained resistivity was more than 30 Ohm-m at this depth.

The resistivity values of subsoil along borehole locations were measured from RI profiles using EarthImager 2D software. The objective of 1D resistivity quantification was to evaluate the predictive capability of the proposed model with soil test results. The extracted resistivity along borehole locations is presented in Figure 4.14.

Once the temperatures at different depths were determined, the observed resistivity at a specific depth was corrected at 15.5°C. Then, the MLR models were used to predict degrees of saturation, using corrected resistivity and CEC at different depths. The comparison of observed degrees of saturation from laboratory experiments at different depths with model-predicted results are presented in Figure 4.15.

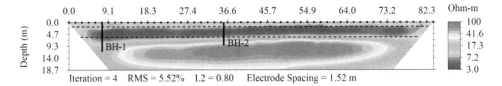

Iteration = 4   RMS = 5.52%   L2 = 0.80   Electrode Spacing = 1.52 m

Figure 4.13 RI at the crest of the slope along highway Loop 12 (Kibria and Hossain, 2016), with permission from ASCE.

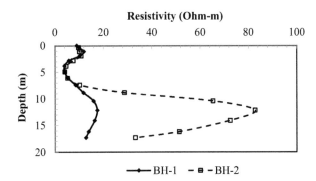

Figure 4.14 Resistivity along boreholes in Loop 12 site

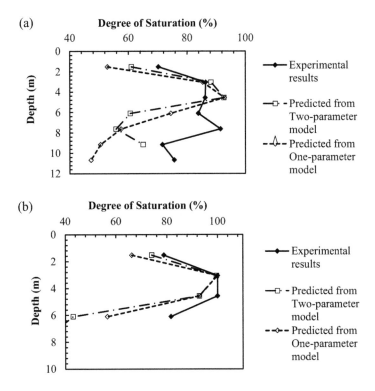

Figure 4.15 Comparison of model-predicted results with experimental observations in Loop 12: (a) BH-1 and (b) BH-2 (Kibria and Hossain, 2016), with permission from ASCE

The comparison between the model-predicted and observed degrees of saturation indicated that the percentage of estimating error was less than 10% up to a depth of 4.6 m for BH-1 and BH-2 when the two-parameter model was utilized. Therefore, the model predictions were in good agreement with the field results at shallow depths. However, a substantial increase in error was identified at depths below 4.6 m. The error was as much as 17.4% at 4.6 m depths in BH-1; nonetheless, a maximum of 12.6% variation was observed in BH-2 when the one-parameter model was used for prediction. The performance of the two-parameter model was better than the one-parameter model.

4.4.3.2.2 RI AT US 287 SOUTH

Hossain (2012) conducted a study on the determination of the active zone in slopes, using real time moisture and matric suction data. Moisture sensors were installed at a depth of 1.22 m at the crest and 2.44 m at the toe of the slope to compare the predictive capability of the MLR models. Soil samples were collected during the test borings to determine the CEC. The CECs of the soil samples collected from 1.22 and 2.44 m depths were 27.5 and 30.6 cmol+/kg, respectively.

RI tests were carried out along sensor locations at the crest and toe of the slope from October 2012 to July 2013. Moisture sensor data was collected, and ambient temperatures were recorded. The RI tests were conducted using dipole-dipole array, with the spacing ranging

Figure 4.16 (a) Location of moisture sensor and RI layout, (b) RI at the crest, (c) RI at toe, and (d) collection of moisture sensor data

from 0.91 to 1.52 m. The apparent resistivity was analyzed using EarthImager 2D software. A finite element method with Cholesky decomposition equation solver and Dirichlet boundary condition was adopted for the forward modeling. Moreover, a high stabilizing (1000) and damping factor (1000) were used in the modeling because traffic movement could cause potential noisy data during the measurements. The locations of the moisture sensor in slopes and the RI tests are presented in Figure 4.16.

4.4.3.2.2.1 Prediction of the degree of saturation at the crest of the slope

The RI results at the crest of the slope from October 2012 to July 2013 are presented in Figure 4.17. A low resistivity zone was identified between approximate depths of 1 to 4 m. The resistivity was as low as 3 Ohm-m at 5.5 m depth. The variations of resistivity over depths along the locations of the moisture sensor were determined using EarthImager 2D and inverted resistivity sections. Based on the profile, resistivity was identified at 1.22 m depth. The resistivity profiles along the sensor locations are presented in Figure 4.18. The observed resistivity was below 15 Ohm-m; however, an increase in resistivity was observed at 0.8 m depth on October 2012.

The temperature at 1.22 m depth was determined, and the observed resistivity was corrected accordingly. The resistivity and CEC (27.5 cmol+/kg) were utilized in the one- and two-parameter MLR models to determine degrees of saturation. The predicted saturations were compared with the sensor results. It should be noted that the volumetric moisture content was obtained from moisture sensors, and degrees of saturation were calculated from the observed volumetric moisture content.

Both the one- and two-parameter models predicted degrees of saturation close to the field results; the maximum prediction errors were 7.47% and 11.17%. A summary of the comparison between the model-predicted and observed degree of saturation is presented in Table 4.3.

4.4.3.2.2.2 Prediction of degree of saturation at the toe of the slope

RI tests were performed at the toe of the slope from October 2012 to July 2013, using dipole-dipole array. The depth of investigation was more than 2.4 m; therefore, a total of 28 electrodes were used. RI results indicated that the resistivity ranged from 7.5 to 12.7 Ohm-m at 2.4. The RI results at the toe of the slope are presented in Figure 4.19.

The resistivity profiles along the sensor location were also identified, as illustrated in Figure 4.20, and the variations of subsurface temperatures were evaluated, using equation 4.14. Once, the resistivity at 2.4 m depth was corrected for temperature, both one-and-two -parameter models were used for the prediction of degrees of saturation.

A summary of the comparison of model-predicted and observed degrees of saturation revealed that the percentage of error was as high as 19.77% in the two-parameter model. The one-parameter model showed a maximum 16.28% error in prediction, as presented in Table 4.4.

4.4.3.3 Discussion of the comparisons

The variations in resistivity can be explained by the anisotropy of the electrical field in the 2D resistivity survey. According to Abu Hassanein *et al.* (1996), the field measurements of electrical resistivity may vary from laboratory test results because of the effect of 3D or 2D electrical fields in the RI tests, the spatial variability of anisotropy and its influence on the results, spatial variations in electrical properties in the subsurface condition, the influence of

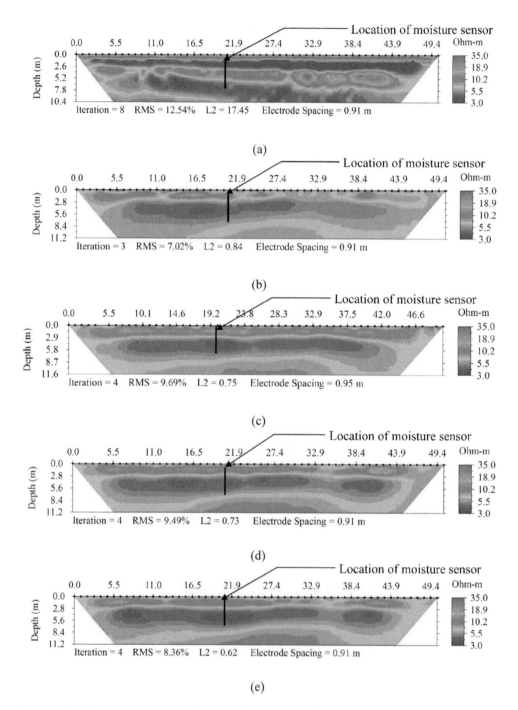

Figure 4.17 RI test results from October 2012 to July 2013 at the crest of slope along US 287: (a) October 2012, (b) November 2012, (c) March 2013, (d) May 2013, and (e) July 2013 (Kibria and Hossain, 2016), with permission from ASCE

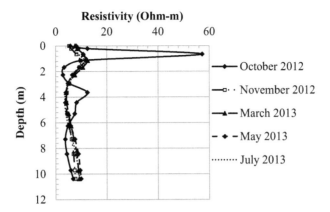

Figure 4.18 Variations of resistivity over depths along sensor locations

Table 4.3 Summary of comparison between predicted and observed degree of saturation at crest (Kibria and Hossain, 2016), with permission from ASCE

Date of RI Tests	Corrected Resistivity (Ohm-m)	Observed Degree of Saturation	Predicted Degree of Saturation (Two-Parameter Model)	Variability (%)	Predicted Degree of Saturation (One-Parameter Model)	Variation (%)
12–Oct–12	9.38	66.41	69.36	−2.95	65.97	0.44
20–Nov–12	9.92	72.50	66.80	5.70	63.37	9.13
24–Mar–13	9.78	62.17	67.45	−5.29	64.03	−1.86
22–May–13	11.81	52.42	59.60	−7.18	56.04	−3.63
8–Jul–13	14.97	58.65	51.18	7.47	47.49	11.17

anomalies in the electrical properties in the subsoils, and the influence of the boundary condition. Cobbles were encountered at shallow depths during instrumentation, and the variations in hydraulic and electrochemical properties in the subsoil condition might have caused the observed variation. In addition, the degree of anisotropy in the 2D electrical field might have increased with the increase in depth.

4.4.4 Evaluation of undisturbed clay properties using the Kibria and Hossain (2016) model

Kibria and Hossain's (2016) undisturbed clay model was validated in the laboratory and in two sites, using borehole data and moisture sensor results. It was evident that the models provided satisfactory results up to about 4.6 m (15 ft.) below the ground surface. Thus, the models are applicable at shallow depths of the field condition and can be used to evaluate geotechnical parameters within the upper 4.6 m (15 ft.).

A set of parameters was evaluated from the Kibria and Hossain (2016) two-parameter undisturbed clay model. It should be noted that the provided geotechnical parameters are applicable within the ranges of parameters used in the model development and should be used for the evaluation of geotechnical parameters in the upper 4.6 m (15 ft.). Similar to the compacted clay model, the cation exchange capacity (CEC) was empirically estimated from

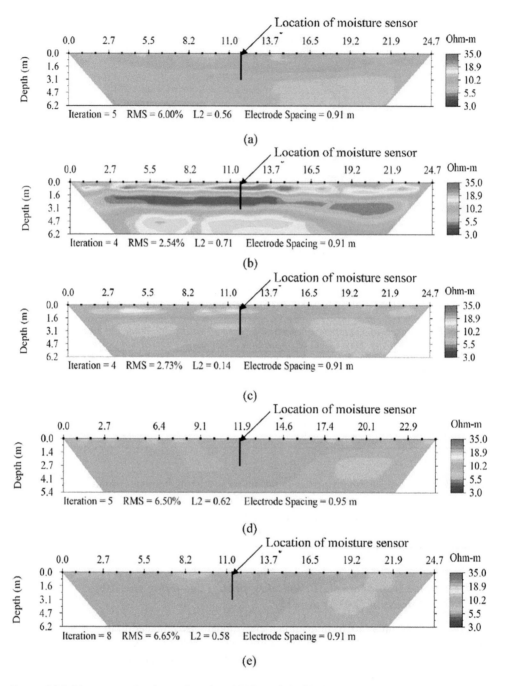

Figure 4.19 RI test results from October 2012 to July 2013 at the toe of slope along US 287: (a) October 2012, (b) November 2012, (c) March 2013, (d) May 2013, and (e) July 2013 (Kibria and Hossain, 2016), with permission from ASCE

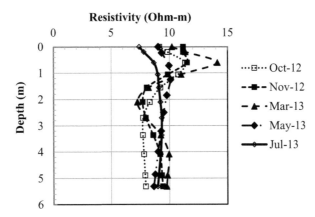

Figure 4.20 Resistivity profile along sensor locations at the toe of slope along US 287

Table 4.4 Summary of comparison between predicted and observed degree of saturation at toe (Kibria and Hossain, 2016), with permission from ASCE

Date of RI Tests	Corrected Resistivity (Ohm-m)	Observed Degree of saturation	Predicted Degree of Saturation (Two-Parameter Model)	Variation (%)	Predicted Degree of Saturation (One-Parameter Model)	Variation (%)
12–Oct–12	9.88	72.87	67.01	5.86	63.58	9.29
20–Nov–12	8.23	72.72	75.64	−2.92	72.35	0.36
24–Mar–13	8.09	60.30	76.54	−16.25	73.27	−12.98
22–May–13	10.83	43.29	63.06	−19.77	59.57	−16.28
8–Jul–13	12.73	43.44	56.77	−13.33	53.17	−9.73

LL, using the correlation proposed by Yukselen and Kaya (2006), and the field resistivity should be corrected for 15.5°C temperature, according to ASTM G187.

4.4.4.1 Evaluation of moisture content

Degree of saturation is a function of moisture content. If the dry unit weight, degree of saturation, and specific gravity of soil are known, then moisture content can be estimated by using the phase relationship of the soil. Thus, Kibria and Hossain's (2016) undisturbed clay model can be used to determine the existing moisture content of the upper 4.6 m (15 ft.) of the soil, according to the following equation:

$$w = \frac{\left\{R^{-0.75} + 0.13063 - 0.00387\left(0.2027LL + 16.231\right)\right\}}{0.00304}\left(\frac{Gs\Upsilon w}{\Upsilon d} - 1\right)\frac{1}{Gs} \tag{4.19}$$

The following example shows the analytical approach to determining moisture content of an existing site, using resistivity data. If the observed resistivity from a RI test in a typical fat clay (CH) soil is 10.0 Ohm-m at 1.5 m depth, the liquid limit (LL) of the soil is 75, the approximate dry unit weight is 14.9 kN/m³, and the specific gravity is 2.65, the estimated *in-situ* moisture content is:

$$w = \frac{\left\{10^{-0.75} + 0.13063 - 0.00387\left(0.2027x75 + 16.231\right)\right\}}{0.00304}\left(\frac{2.65x9.81}{14.9} - 1\right)\frac{1}{2.65}$$

$$w = 17.3\%$$

The moisture content obtained from the equation is in good agreement with the correlation chart presented in Figures B1 to B4 in Appendix B, applicable for this sample.

4.4.4.2 Evaluation of dry unit weight

According to the phase relationship of soil, degree of saturation is a function of dry unit weight. At a given degree of saturation, specific gravity, moisture content, and dry unit weight of an existing site can be determined analytically from resistivity by using Kibria and Hossain's (2016) model in accordance with the following equation:

$$\left(\frac{1}{\Upsilon d}\right) = \left[\frac{0.00304w.Gs}{\left\{R^{-0.75} + 0.13063 - 0.00387\left(0.2027LL + 16.231\right)\right\}} + 1\right].\frac{1}{\Upsilon wGs} \qquad (4.20)$$

The following is an example of how to determine the dry unit weight of soils by using resistivity. If the observed resistivity from a RI test in a typical fat clay (CH) soil is 10.0 Ohm-m at 1.5 m depth, the LL of the soil is 75, the observed moisture content from the geotechnical investigation is 10%, and the specific gravity is 2.65, the estimated dry unit weight of that depth is:

$$\left(\frac{1}{\Upsilon d}\right) = \left[\frac{0.00304x10x2.65}{\left\{10^{-0.75} + 0.13063 - 0.00387\left(0.2027x75 + 16.231\right)\right\}} + 1\right].\frac{1}{9.81x2.65}$$

$$\Upsilon d = 18.2 \text{kN} / \text{m}^3$$

According to the presented equation, the unit weight of the sample is 18.2 kN/m³. The estimated unit weight from the equation is in agreement with the correlation chart presented in Figures B5 to B7 in Appendix B, applicable for this sample.

4.4.4.3 Evaluation of degree of saturation

Kibria and Hossain's (2016) undisturbed clay model was correlated with the degree of saturation. Therefore, at a known LL, the degree of saturation can be estimated by using the electrical resistivity of soil. The following equation was derived from Kibria and Hossain's (2016) two-parameter undisturbed clay model to determine the degree of saturation:

$$S_R = \frac{R^{-0.75} + 0.13063 - 0.00387\left(0.2027LL + 16.231\right)}{0.00304} \qquad (4.12)$$

The following example shows how to determine the degree of saturation of undisturbed soils by using resistivity. If the observed resistivity from a RI test in a typical fat clay (CH) soil is 10.0 Ohm-m at 1.5 m depth, and the LL and specific gravity of the soils are 75 and 2.65, respectively, then the degree of saturation of the sample is:

$$S_R = \frac{10^{-0.75} + 0.13063 - 0.00387(0.2027 x75 + 16.231)}{0.00304}$$

$$S_R = 61.4\%$$

The equation indicates that the degree of saturation of the sample is 61.4%.; therefore, the estimated degree of saturation is in agreement with the correlation chart presented in Figure B8 in Appendix B and is applicable for this sample.

4.4.4.4 Evaluation of void ratio

Degree of saturation is a function of void ratio, according to phase relation of soils. For a given degree of saturation and specific gravity, the void ratio can be back-calculated. This analytical approach can be used to determine the void ratio from resistivity of undisturbed soils using Kibria and Hossain (2016). Since it is possible to estimate LL, using CEC, it is possible to determine the void ratio from resistivity at a specific moisture content by using the following equation:

$$e = \frac{0.00304 w.Gs}{\left\{R^{-0.75} + 0.13063 - 0.00387(0.2027 LL + 16.231)\right\}} \tag{4.22}$$

The following is an example of how to estimate the void ratio of soils, using resistivity. If the observed resistivity of a typical undisturbed fat clay (CH) soil is 10.0 Ohm-m at 1.5 m depth, the liquid limit (LL) of the soil is 75, the observed moisture content from the geotechnical investigation is 10%, and the specific gravity is 2.65, the void ratio of the sample is:

$$e = \frac{0.00304 x10 x2.65}{\left\{10^{-0.75} + 0.13063 - 0.00387(0.2027 x75 + 16.231)\right\}}$$

$$e = 0.43$$

The equation indicates that the void ratio of the sample is 0.43; therefore, the estimated void ratio is in agreement with the design chart presented in Figures B9 to B11 in Appendix B, and is applicable for this sample.

4.4.4.5 Evaluation of existing compaction condition

The existing compaction condition is important for determining the geohazard potential. If the dry unit weight is determined using resistivity, and the maximum dry density of the soil sample is known, the existing compaction level can be evaluated according to the equations presented below:

$$\left(\frac{1}{\Upsilon d}\right) = \left[\frac{0.00304w.Gs}{\left\{R^{-0.75} + 0.13063 - 0.00387\left(0.2027LL + 16.231\right)\right\}} + 1\right].\frac{1}{\Upsilon wGs} \qquad (4.23)$$

$$Compaction\ Level = \left(\frac{\Upsilon d}{\Upsilon max}\right) \qquad (4.24)$$

The following is an example of how to estimate the existing compaction condition of soils, using resistivity. If the observed resistivity from a RI test in a typical fat clay (CH) soil is 10.0 Ohm-m at 1.5 m depth, the liquid limit (LL) of the soil is 75, and the assumed optimum moisture content (OMC) and maximum dry density of the sample are 18% and 16.5 kN/m³, respectively, the compaction level of the sample is:

$$\left(\frac{1}{\Upsilon d}\right) = \left[\frac{0.00304x18x2.65}{\left\{10^{-0.75} + 0.13063 - 0.00387\left(0.2027x75 + 16.231\right)\right\}} + 1\right].\frac{1}{9.81x2.65}$$

$\Upsilon d = 14.6 kN/m3$
$Compaction\ Level = \left(\dfrac{14.6}{16.5}\right) = 0.88$

The observed compaction level is 0.99; therefore, the obtained compaction level is in agreement with the design chart presented in Figures B12 to B17 in Appendix B and is applicable for this sample.

4.4.5 Limitations of the Kibria and Hossain models (2015 and 2016)

It should be noted that the practically applicable models, evaluated geotechnical properties, and proposed correlations charts presented herein are a non-invasive way to approximate geotechnical parameters from resistivity imaging. The proposed correlations should be considered approximate and should not be used as an alternative to soil test borings and laboratory testing. Since resistivity is a complex phenomenon and depends on many factors, the proposed correlations are subject to change under various circumstances, and the evaluated geotechnical properties should be verified in the field.

4.4.6 Evaluation of corrosion potential (Kibria and Hossain, 2017)

Electrical resistivity has been widely used in the evaluation of the corrosion potential of soils. Caltrans Corrosion Guidelines (Caltrans, 2003) and American Water Works Association (AWWA, 1999) indicate that resistivity is an important factor for evaluating the corrosion potential of soils. The correlations of corrosion exposure level and ranges of resistivity are presented in Table 4.5.

Even though the determination of electrical resistivity is important for the evaluation of corrosion potential, it is often impossible to measure resistivity in the laboratory or field. Thus, an approximate estimation of corrosion potential from geotechnical properties is vital for project planning purposes.

Kibria and Hossain (2017) developed correlations of LL, PI, activity, and CEC with electrical resistivity of soil. The proposed correlations provided a mean to evaluate corrosion potential

Table 4.5 Ranges of electrical resistivity and associated corrosion potential (Robinson, 1993)[10]Reproduced with permission from NACE International, Houston, TX. All rights reserved. Robinson W., Testing Soil for Corrosiveness, Materials Performance, Vol. 32, Issue 4, 1993. © NACE International 1993.

Soil Resistivity (Ohm-cm)	Corrosivity	Designated Exposure Type
>25000	Relatively corrosive	I
10000–25000	Slightly corrosive	II
3000–10000	Moderately corrosive	III
1000–3000	Corrosive	IV
500–1000	Very corrosive	V
0–500	Severely corrosive	VI

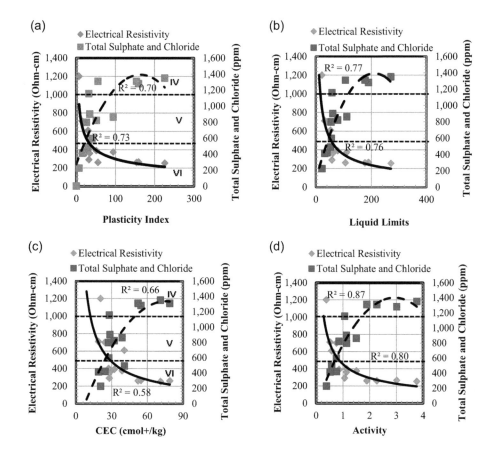

Figure 4.21 Correlation of electrical resistivity and total sulfate and chloride ions with geotechnical properties (Kibria and Hossain, 2017), with permission from ASCE

of soils. Since the index properties are very well known to geotechnical engineers, the corrosive exposure level can be approximated using the method from physical properties of soil.

The variations of saturated minimum resistivity with the estimated total sulfate and chloride content of test specimens with LL, PI, CEC, and the activity of the test specimens are presented in Figure 4.21.

Test results indicated that an increase LL, PI, CEC, and activity caused reductions in saturated minimum resistivity values, but increases in sulfate and chloride content. Since resistivity, sulfate, and chloride content are correlated with LL, PI, activity, and CEC, the corrosion potential can be approximated by using these geotechnical properties. A flow chart of a corrosion exposure evaluation using geotechnical properties is presented in Figure 4.22.

Six artificially prepared test specimens were used to validate the efficiency of the proposed method of evaluating corrosion potential. The geotechnical properties and saturated minimum resistivity of the samples were determined in the laboratory. Additionally, the resistivity was estimated using Figure 4.21.The observed variations are presented in Table 4.6.

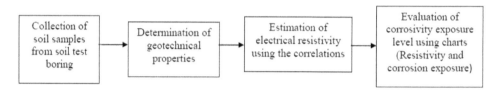

Figure 4.22 Steps for estimation of corrosion potential of subsoil using geotechnical properties (Kibria and Hossain, 2017), with permission from ASCE

Table 4.6 Comparison of predicted and measured resistivity for the evaluation of corrosion potential (Kibria and Hossain, 2017), with permission from ASCE

Sample Designation	Sample Description	LL	PI	Activity	Estimated Saturated Minimum Resistivity (Average) (Ohm-cm)	Measured Saturated Minimum Resistivity (Ohm-cm)	Variation (%)
A	Illite 100%	50	23	0.23	740	737	−0.5
B	Ca ben–20%–Illite 40%–Kaolin 20%– Sand 20%	54	24	0.31	667	815	18.2
C	Ca ben–40%–Illite 20%–Kaolin 20%– Sand 20%	65	30	0.38	594	523	−13.6
D	Na ben–20%–Illite 40%–Kaolin 20%– Sand 20%	86	59	0.74	436	490	11.1
E	Na ben–40%–Illite 20%–Kaolin 20%– Sand 20%	131	100	1.25	332	365	8.9
F	Na ben–20%–Illite 20%–Kaolin 20%– Sand 40%	76	55	0.92	426	417	−2.0

Although estimated and measured resistivity varied as much as 18.2%5, the corrosion exposure levels of the samples were identical in each case when compared to estimated resistivity and measured resistivity. Since significant uncertainties are involved with corrosion parameters, Kibria and Hossain (2017) emphasized that the estimated corrosion exposure level should be considered approximate and needs to be verified in the field.

Chapter 5

Electrical resistivity of municipal solid waste (MSW)

5.1 General

Over the last few decades, the generation, recycling, and disposal of municipal solid waste (MSW) have changed substantially. In the United States, MSW generation increased from 3.66 to 4.34 pounds per person per day between 1980 and 2009. About 243 million tons of MSW was generated in 2009 (EPA, 2009). Approximately 33.8% of MSW was recycled and composted, 11.9% was converted to energy, and 54.3% (about 132 million tons) was discarded in landfills. In the foreseeable future, landfilling will remain a major solid waste disposal method for MSW.

In conventional landfills that are designed and operated in accordance with Subtitle D of the Resource Conservation and Recovery Act (RCRA), efforts are typically made to minimize the moisture entering the landfill so that the generation of leachate is minimized and the risk of groundwater contamination is reduced. However, the time required for the decomposition of waste in a dry tomb landfill typically ranges from 30 to 100 years, and the landfill gas is expected to be produced at a slow rate over a long period of time.

In the mid-1970s, Pohland (1975) proposed the idea of enhancing waste decomposition by recirculating the leachate and/or adding water. Additional moisture stimulates microbial activity by providing better contact between insoluble substrates, soluble nutrients, and microorganisms (Barlaz *et al.*, 1990). As a result, decomposition and biological stabilization of MSW can be reduced to years, as compared to decades, for traditional dry landfills. The fundamental aspect of the operation of a bioreactor landfill is the controlled addition of water and/or the recirculation of the generated leachate back into the landfill's waste mass. Several studies have pointed out the potential benefits of the bioreactor landfill approach (Barlaz *et al.*, 1990; Reinhart and Townsend, 1998; Pacey, 1999; Warith, 2002), which include (1) increased rate of settlement of MSW which results in increasing the landfill's capacity, (2) increased rate of landfill gas production for energy recovery projects, (3) stabilization of waste in a shorter period of time, reduction of the post-closure monitoring costs, and (4) reduction of leachate treatment and disposal costs. As a result of these benefits, there has been an increasing trend to operate landfills as bioreactor landfills, particularly in areas where landfill space is an important issue.

Since there is an increasing trend to operate landfills as bioreactor landfills, it is crucial to understand and monitor the moisture content and distribution within the landfill. Monitoring the moisture distribution within a bioreactor landfill is essential not only for the design and operation of the leachate recirculation systems, but also for identifying sites with non-uniform leachate distribution due to ponding and channeling. Several methods have been developed and implemented to measure the moisture content of MSW (Imhoff *et al.*, 2007). The most common methods are: (1) waste sampling using drilling, (2) moisture sensors, or (3) probe measurements, (Figure 5.1), all of which are intrusive methods and provide data for localized waste. Although the direct method of waste sampling and determining the moisture content gravimetrically provides an accurate measurement of waste moisture content, it is expensive to sample the waste and is intrusive to the containment of the waste. In addition, a large number of MSW samples are necessary for accurate determination of moisture distribution because of the waste heterogeneity.

Moisture measuring probes, such as time domain reflectometry (TDR) probes and sensors, are commonly used to provide waste moisture. One of the limitations associated with using such probes is the poor contact between the probes or sensors and waste. With time and solid waste decomposition, the moisture content of MSW is expected to change significantly. Also, solid waste materials are expected to settle considerably with time. Therefore, the possibility of sensors being lost or short circuited during the leachate recirculation period is very high, resulting in poor moisture content readings. The use of sensors is highly invasive and requires a lot of instrumentation. For example, a study conducted at the New River Regional Landfill in Florida, using the resistance-based sensors method, required approximately 65,000 ft. of wiring. All of these methods provide localized information of moisture content, not a general view of the entire site.

Over the last decades, the electrical resistivity technique has improved significantly for geotechnical and geoenvironmental site investigations. The resistivity imaging (RI) test has been a very popular site investigation and characterization tool for different geotechnical and geoenvironmental applications over the years (Kalinski and Kelly, 1993; Dahlin, 2001, Khan *et al.*, 2012). Different factors affecting the resistivity of the soil and solid waste were studied by several researchers, and the results indicated that electrical resistivity varies with the water content, temperature, ion content, particle size, resistivity of the solid phase, permeability, porosity, clay content, degree of saturation, organic content, and pore water composition present in the materials (McCarter, 1984; Abu Hassanein *et al*, 1996; Giao *et al.*, 2003; Grellier *et al.*, 2006; Samouëlian *et al.*, 2005; Kibria and Hossain, 2012; Clement *et al.*, 2010; Ekwue and Bartholomew, 2010). Shihada *et al.* (2013)

Figure 5.1 Moisture measurement at the MSW landfill: (a) Drilling with 3 ft. dia. bucket auger sampling (b) Resistance sensors study at New River Regional Landfill (Reinhart and Townsend, 2007)

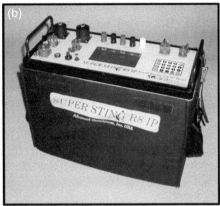

Figure 5.2 (a) Resistivity test box and (b) Super Sting resistivity meter (Shihada, 2011)

conducted a laboratory study that indicated that the electrical resistivity of solid waste is a complex property that depends on the moisture content, composition, unit weight, pore fluid composition, temperature, and decomposition and organic content. However, the changes of the resistivity highly depend on the moisture content and temperature, and the effects of the other parameters on resistivity have not been well established for solid waste (Grellier *et al.*, 2007)

Shihada (2011) conducted a study using the electrical resistivity of fresh MSW samples, landfilled MSW samples, and degraded MSW. An acrylic rectangular box (Figure 5.2) was designed, based on the four-electrode method, to measure the electrical resistivity of MSW. It had two metal ends and two pins inserted across its length. The dimensions of the box were 14.8 cm x 15.5 cm x 29.5 cm. The resistivity of the MSW sample was measured by passing a current between the ends and measuring the voltage drop between the interior pins. The basic equation associated with the resistivity box is:

$$\rho = \frac{WH}{L} R \tag{5.1}$$

where ρ = specimen resistivity; W = width of the box; H = height of box; R = measured resistance; and L = distance between the inner pins.

Shihada (2011) also investigated the effects of moisture content, unit weight, decomposition, temperature, composition of MSW, and composition of pore fluid on resistivity of MSW. The study is summarized in the following section.

5.2 Effect of moisture content on electrical resistivity

Shihada (2011) conducted electrical resistivity on five fresh MSW samples, 30 landfilled MSW samples, and four degraded MSW samples at their actual moisture contents. The fresh MSW samples were collected from the working face of an MSW landfill, and the 30 landfilled samples were collected from different depths of a bioreactor cell of MSW landfill.

Four laboratory-scale bioreactors were formed to collect MSW samples at different phases of decomposition. Taufiq (2010) conducted a study on the effects of different compositions of MSW on the unit weight. Using the standard Proctor compaction method, the study determined that the moist unit weight of fresh MSW is approximately 35 lbs./ft³. Oweis and Khera (1998) presented that due to moderate-to-good compaction efforts, the unit weight of soil ranges between 30–45 lbs./ft³. Shihada (2011) compacted the samples, using the standard compaction test to a moist unit weight of 35 lbs./ft³. Each of the fresh MSW samples was dried in an oven at 105°C for 24 hours, then was compacted to a dry unit weight of 35 lbs./ft³. The remolded sample was then transferred to the resistivity box. Tap water (resistivity ranging from 30–32 ohm-m) was added daily to the sample, and the corresponding moisture content was determined. The sample was covered with a plastic wrap to prevent moisture loss and remained in the test box for at least 24 hours to ensure even moisture distribution before the resistivity tests were conducted.

Shihada (2011) conducted electrical resistivity tests on three landfilled MSW samples collected from different depths (20, 30, and 40 ft.) of three boreholes designated as B45, B47, and B49. The waste composition of the samples is presented in Figure 5.3. Each of the landfilled MSW samples was oven dried and compacted to a dry unit weight of 21 lbs./ft³.

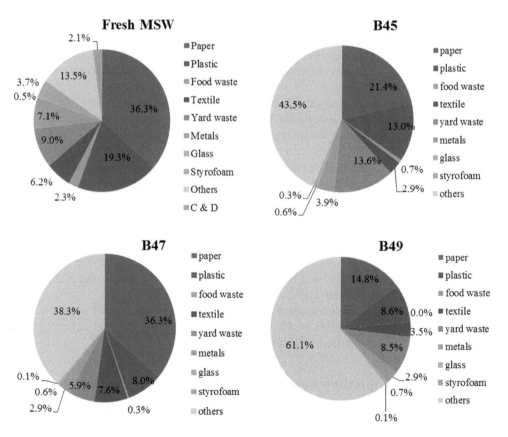

Figure 5.3 Waste composition for electrical resistivity imaging test (Shihada *et al.*, 2013), with permission from ASCE

The remolded sample was transferred to the resistivity box, and tap water was added to the sample to vary the moisture content. The sample was left in the test box for at least 24 hours to ensure even moisture distribution, and then the resistivity tests were conducted.

Shihada (2011) conducted further resistivity tests on four degraded MSW samples that were collected from laboratory-scale bioreactors. Each of the degraded MSW samples was oven dried and compacted, using the same compaction effort (25 blows with a standard Proctor test hammer on each of the four layers). When necessary, Shihada (2011) also utilized a universal testing machine to compact the samples when the target unit weight could not be attained by using the standard Proctor compaction effort. The remolded sample was then transferred to the resistivity box, and tap water was added to the sample to vary the moisture content. The sample was left in the test box for at least 24 hours to ensure even moisture distribution, and then the resistivity tests were conducted. The sample preparation procedure of the study is presented in Figure 5.4.

5.2.1 Fresh MSW samples

Electrical resistivity is the inverse of conductivity, which decreases with an increase in moisture content. The conduction in MSW is largely electrolytic, and the electrical current is carried by the ions in the pore fluid. More ions in the pore fluid result in increased conductivity. Based on the study conducted by Shihada (2011), the electrical resistivity of the five fresh

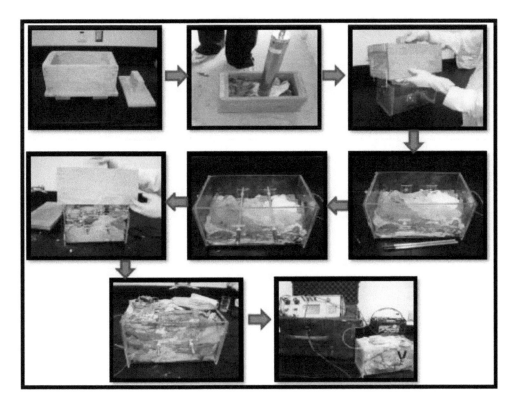

Figure 5.4 Sample preparation procedure for resistivity test (Shihada, 2011)

Figure 5.4 Continued

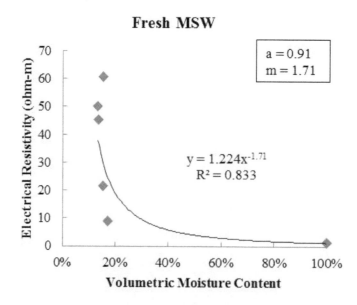

Figure 5.5 Variations of electrical resistivity at different moisture contents of fresh MSW (Shihada *et al.*, 2013), with permission from ASCE

MSW samples were plotted against their corresponding field moisture contents, as presented in Figure 5.5, where a decrease in electrical resistivity with an increase in moisture content can be observed. The resistivity of the fresh MSW samples at their field moisture contents ranged from approximately 8–60 ohm-m when compacted to a unit weight of 35 lbs./ft³ and was within a reasonable range compared to other studies. Grellier et al., (2007) indicated an electrical resistivity range of 5 to 100 ohm-m, and Imhoff *et al.* (2007) indicated a range of 5 to 85 ohm-m for MSW at different moisture contents.

Shihada (2011) fitted the experimental data with Archie's law and the determined constants. Archie's law constant, 'a', varied from 0.87 to 0.96, and the constant 'm' varied from 1.0 to 1.61 for five different samples when analyzed individually. The variations of the electrical resistivity took place due to the change in composition, temperature, and pore fluid mineralization. Due to mineralization, the resistivity of the water dropped from the 30–32 ohm-m range to the 1.14–1.72 ohm-m range. The combined test data of all five samples utilized by Shihada (2011) is presented in Figure 5.5, which fits reasonably with Archie's law with the constants a = 0.91 and m = 1.45. An average pore fluid resistivity of 1.35 ohm-m was used to determine these constants.

5.2.2 Landfilled MSW samples

Shihada (2011) utilized landfilled MSW samples to investigate the changes in the electrical resistivity with varying moisture content. The electrical resistivity of 30 landfilled MSW samples was measured at actual field moisture content. Figure 5.6 presents a plot of electrical

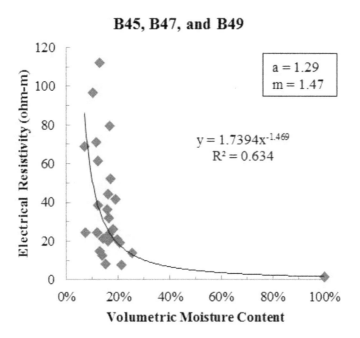

Figure 5.6 Variation of electrical resistivity with different field moisture content of landfilled MSW samples (Shihada *et al.*, 2013), with permission from ASCE

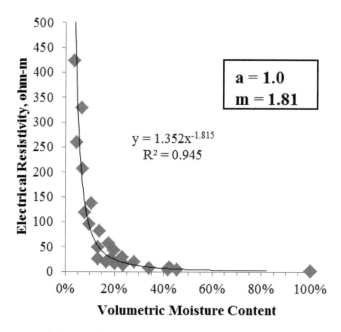

Figure 5.7 Variation of electrical resistivity with different moisture content using tap water of landfilled MSW samples (Shihada *et al.*, 2013), with permission from ASCE

resistivity versus the corresponding field moisture content for all 30 landfilled samples. Similar to fresh MSW samples, the electrical resistivity decreased with an increase in moisture content. The resistivity of the landfilled samples at their field moisture content ranged from approximately 8–112 ohm-m when compacted to a unit weight of 35 lbs./ft³. Archie's law was fitted to the experimental data, and the constants were determined. The constant 'a' was found to be 1.29 and 'm' was 1.47, which are in the same order as soil and rocks.

To better understand the effects of moisture content on resistivity, Shihada (2011) conducted further resistivity tests on three landfilled MSW samples which were collected from a 3 ft. diameter auger borehole at depths of 20, 30, and 40 ft. The landfilled samples were completely dried at the beginning, and then tap water was added to the sample to vary the moisture content. The results, presented in Figure 5.7, show that the effect of increasing moisture content on electrical resistivity tapered off beyond a certain point (moisture content of approximately 50% to 55%). This can possibly be explained by the fact that at higher values of moisture content, continuous current flow paths through the pores would have already been established. Archie's law constants were determined to be 'a' = 1.0 and 'm' = 1.81.

5.2.3 Degraded MSW samples

Shihada (2011) conducted electrical resistivity tests on four degraded MSW samples by drying each sample and adding tap water to the sample to vary the moisture content. The dried samples were compacted by applying 25 blows on each of four layers, using a standard Proctor test hammer. The test results, presented in Figure 5.8, show that the four samples had

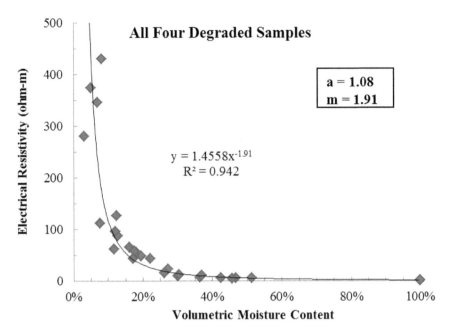

Figure 5.8 Electrical resistivity variation with moisture content of all degraded samples (Shihada *et al.*, 2013), with permission from ASCE

Table 5.1 Archie's law constants for fresh, landfilled, and degraded MSW (Shihada *et al.*, 2013), with permission from ASCE

Sample	A	m
Fresh MSW	0.91	1.45
Landfilled MSW	1.0	1.81
Degraded MSW	1.08	1.91

similar trends to those of fresh MSW and borehole samples. Archie's law was fitted to the experimental results. Archie's law constants were determined for all degraded samples to be 'a' = 1.08 and 'm' = 1.91.

The Archie's law constants for fresh, landfilled, and degraded MSW samples are presented in Table 5.1. The constant 'a' varied slightly for fresh, landfilled, and degraded MSW samples. The constant 'm' varied from 1.45 to 1.91. Grellier *et al.* (2007) determined that 'a' was 0.75 and 'm' ranged from 1.6 to 2.15 for MSW samples from the Orchard Hill landfill in Illinois.

These constants are in the same order for both soil and rocks. According to Keller and Frischknecht (1966), the value of the constant 'a' varies from slightly less than one (1) for rocks with intergranular porosity to slightly more than one (1) for rocks with joint porosity. The exponent 'm' is larger than two (2) for cemented and well-sorted granular rocks and less than two (2) for poorly sorted and poorly cemented granular rocks. Jackson *et al.* (1978) found that the exponent 'm' was dependent on the shape of the particles, which increased as they become less spherical, while variations in size and spread of sizes appeared to have little effect. It is difficult to make similar conclusions for MSW due to the heterogeneity of the waste.

5.3 Effect of unit weight

Shihada (2011) investigated the effect of unit weight on electrical resistivity, using five fresh MSW samples, five landfilled MSW samples, and four degraded MSW samples at their actual moisture content and at three different unit weights. Fresh MSW samples at their actual field moisture content were prepared at 35 lbs./ft.^{3}, 45 lbs./ft.^{3}, and 55 lbs./ft.^{3} by increasing the compaction effort. Compaction was done by applying a load ranging from 1600 to 2000 pounds, using a 60 kip tensile-compression machine until the target unit weight was reached. The compacted samples were left in the test box for at least 24 hours, and then the resistivity tests were conducted. The test results are presented in the following section.

5.3.1 Fresh MSW samples

Shihada (2011) compacted five fresh MSW samples at their field moisture contents to a unit weight of 35, 45, and 55 lbs./ft.^{3}. The variation of electrical resistivity with unit weight for the five fresh MSW samples with different moisture contents is presented in Figure 5.9. The results showed that resistivity decreased with an increase in unit weight. As unit weight increases, air voids are reduced, resulting in an increase in the degree of saturation. An increase in saturation means that more voids are filled with liquid, creating more paths for

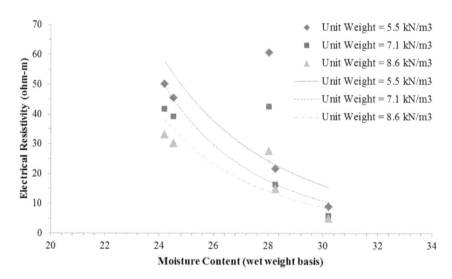

Figure 5.9 Effect of unit weight and moisture content on the electrical resistivity of fresh MSW samples (Shihada *et al.*, 2013), with permission from ASCE

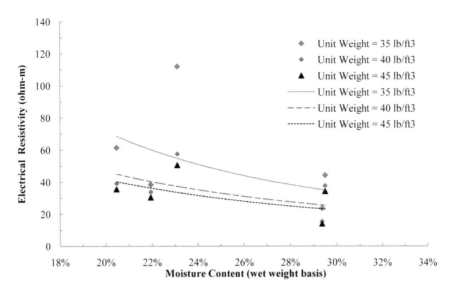

Figure 5.10 Effect of unit weight and moisture content on electrical resisting of landfilled MSW samples. (Shihada, 2011)

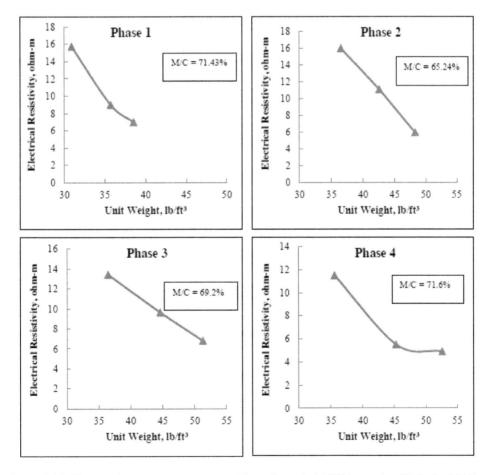

Figure 5.11 Electrical resistivity variation of four degraded MSW samples (Shihada, 2011)

current flow, and therefore decreasing the electrical resistivity. Thus, the resistivity of MSW decreases with increasing moisture content and increasing unit weight.

5.3.2 Landfilled MSW samples

Shihada (2011) investigated five landfilled MSW samples taken from a 3-ft. diameter auger borehole at field moisture contents and compacted to a unit weight of 35, 40, and 45 lbs./ft.[3] The variations of electrical resistivity with unit weight for the five samples are presented in Figure 5.10. It is evident from the results that resistivity decreases with increasing unit weight and increasing moisture content, which is similar to the trend observed for fresh MSW.

5.3.3 Degraded MSW samples

Shihada (2011) obtained four degraded MSW samples at their actual moisture content from laboratory scale bioreactors and compacted them to an approximate unit weight of 30 to 55 lbs./ft.[3], using a standard Proctor test hammer. The variations of electrical resistivity with unit weight for the four degraded samples are presented in Figure 5.11. Again, it is evident from the results that resistivity decreases with an increase in unit weight.

5.4 Effect of decomposition

Shihada investigated the effects of decomposition of the MSW on the samples prepared in the laboratory scale reactors. Electrical resistivity tests were conducted on the four degraded MSW samples at their actual moisture contents upon dismantling the reactors. Reactor samples were compacted using the same compaction effort (25 blows on each of the two layers, using a standard Proctor test hammer). The unit weights of the samples from phases 1, 2, 3, and 4 were 35.71, 48.31, 51.32, and 52.59 lbs./ft[3], respectively, indicating an increase in unit weight with decomposition. This conclusion is consistent with other results found in literature. According to Dixon and Jones (2005), the bulk unit weight of fresh MSW ranges from 6–7 kN/m[3], while that of degraded waste ranges from 14–20 kN/m[3]. Haque (2007) found that the unit weight of MSW increased from 8.5 kN/m[3] in phase 1 of decomposition to 10.7 kN/m[3] in phase 4. This was due to the reduction in particle size with decomposition, resulting in reducing the voids and increasing the mass of solids per unit volume. The resistivity results, as presented in Figure 5.12, indicate that electrical resistivity decreased with decomposition from 8.98 ohm-m in phase 1 to 4.91 ohm-m in phase 4. This decrease in resistivity was probably caused by the increase in unit weight caused by decomposition.

Shihada (2011) conducted further resistivity tests on the four degraded MSW samples by drying each sample and adding tap water to the sample to vary the moisture content. The resistivity results for the four degraded samples are presented in Figure 5.13. As shown in Figure 5.13, there was a clear distinction between the early stage (phase 1) and the late stage (phase 4) of decomposition. There was little difference between the resistivity values of the phase 2 and phase 3 samples. It is important to note that the resistivity values were plotted against the gravimetric moisture content in Figure 5.13. When they were plotted against the volumetric moisture content (Figure 5.14), there was no clear distinction between the phases of decomposition, and most points were fitted to one curve. This is because the volumetric moisture content takes into account the different unit weights of the samples. Therefore, it

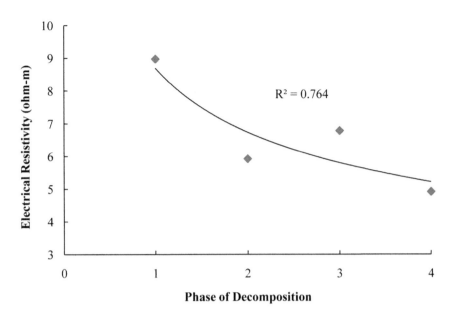

Figure 5.12 Variation of electrical resistivity on the different phase of decomposition of the MSW (Shihada *et al.*, 2013), with permission from ASCE

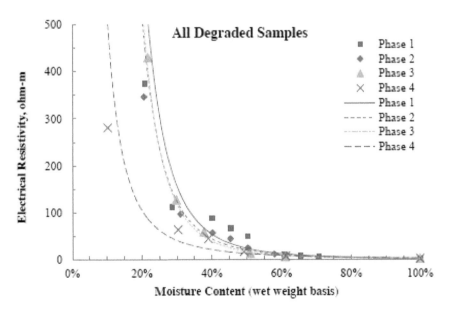

Figure 5.13 Electrical resistivity variations with moisture content with different phases of degradation (Shihada, 2011)

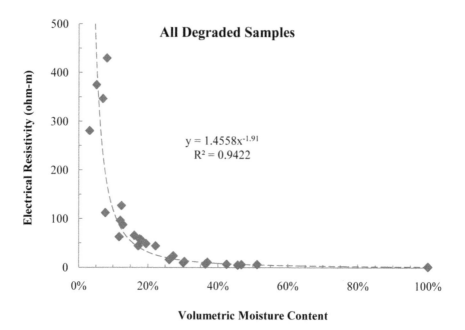

All Degraded Samples

Figure 5.14 Electrical resistivity variations with volumetric moisture content of the degraded samples (Shihada *et al.*, 2013), with permission from ASCE.

can be concluded that the observed decrease in resistivity with decomposition is mainly due to the increase in unit weight that occurs as a result of decomposition.

5.5 Effect of temperature

An increase in temperature decreases the viscosity of water, causing the ions in the water to become more mobile. Thus, the electrical conductivity increases and the resistivity decreases with increasing temperature. In general, electrical resistivity decreases by about 2% for a temperature increase of 1°C. According to Keller and Frischknecht (1966),

$$\rho_t = \frac{\rho_{18}}{1 + \alpha(t - 18)} \tag{5.2}$$

where α is the temperature coefficient ($\alpha \approx 0.025$ per °C), ρt is the resistivity at ambient temperature t(°C), and $\rho 18$ is the resistivity at a reference temperature of 18°C. (Any other reference temperature may be used).

Shihada (2011) conducted all laboratory testing at room temperature (70°F). Before using any correlation that was developed in the laboratory, the resistivity values obtained from field investigation have to be corrected to a standard temperature (70°F). An increase in temperature decreases the viscosity of water, causing the ions in the water to become more mobile. Therefore, the electrical conductivity of MSW is expected to increase, and the resistivity is expected to decrease with increasing temperature.

Temperature affects the electrical resistivity of MSW because it affects the mobility of the ions in the pore fluid. To investigate the effect of temperature, Shihada (2011) investigated the resistivity of five fresh MSW samples at different temperatures. The results for the variations of resistivity with temperature are presented in Figure 5.15. The model suggested by Keller and Frischknecht (1966) (using $\alpha = 0.025$) was also fitted to the data, and the model fit the data well. The temperature coefficient (α) was determined for each sample, and the average temperature coefficient was determined to be 0.020 per degree Celsius, as presented in Table 5.2. These results verify that the Keller and Frischknecht equation (equation 5.2) can

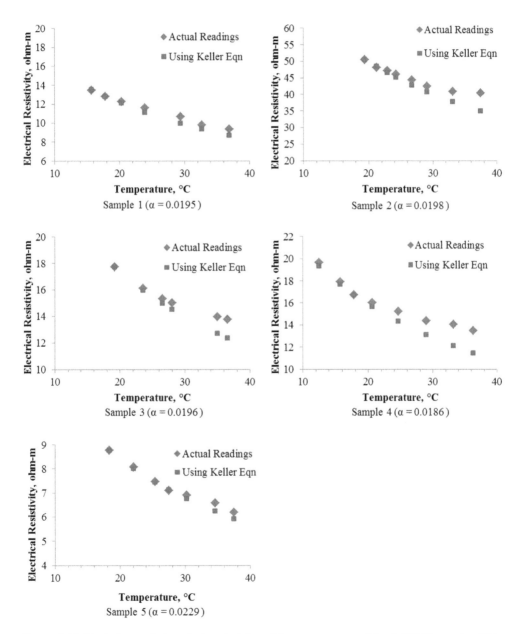

Figure 5.15 Variations of electrical resistivity with temperature of fresh MSW samples (Shihada *et al.*, 2013), with permission from ASCE

Table 5.2 Temperature coefficients for five fresh
MSW samples (Shihada, 2011)

Sample	α *(per °C)*
1	0.0195
2	0.0198
3	0.0196
4	0.0186
5	0.0229
Average	**0.020**

be used to correct field resistivity values to a standard temperature (70°F). The results were consistent with the findings by Grellier *et al.* (2005) that the resistivity of leachate decreases by 2% per temperature increase of 1°C.

5.6 Effect of composition

Electrical conduction in solid waste is largely electrolytic, taking place in the liquid contained in the pores (Guerin *et al.*, 2004). However, the effect of the composition of solid waste on electrical conduction is not clearly understood and has not been studied. In soil science, it is well known that the type of rock affects electrical resistivity. Typical ranges of resistivity for a number of rock and soil types are well documented. Similarly, it is expected that the solid content of the waste may have an effect on electrical resistivity. Shihada (2011) investigated the effect of composition of MSW on electrical resistivity. The study is summarized in the following section.

5.6.1 Fresh MSW samples

Shihada (2011) manually sorted the individual components after conducting the resistivity tests from the samples in the test box. The weight percentages of the paper and other components were determined for five fresh MSW samples and 30 landfilled MSW samples. Paper and "others" were selected because they were the dominant components in the waste samples. The average physical composition by weight of the samples is presented in Figure 5.3.

The resistivity of each sample was plotted against its corresponding paper content and "others" content to find any possible trends. Paper and the "others" components were selected because they were the dominant components in the waste samples.

Figure 5.17 presents a plot of the electrical resistivity versus the corresponding paper percentage by weight for five fresh MSW samples. A decrease in the electrical resistivity with an increase in paper content was observed. This was probably due to the fact that paper products tend to absorb more water, and increases in paper content usually correspond to higher moisture content.

On the other hand, an increase in electrical resistivity was observed when the "others" (soil and fines) content increased, as shown in Figure 5.17. Shihada (2011) explained this behavior by the fact that an increase in the fines content decreases the porosity, reducing the number of voids available for current flow.

5.6.2 Landfilled MSW samples

Shihada (2011) utilized landfilled samples from a 3-ft. diameter auger borehole (designated as B45, B47 and B49), and manually sorted them into individual components. Figure 5.3 depicts

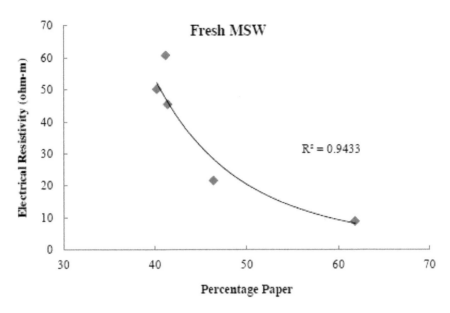

Figure 5.16 Effect of paper content on resistivity of fresh MSW samples (Shihada *et al.*, 2013), with permission from ASCE

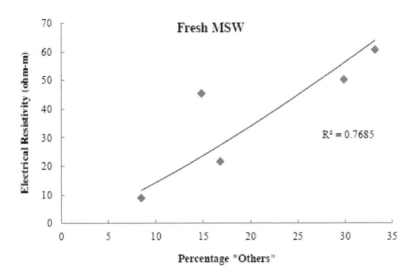

Figure 5.17 Effect of "others" content on electrical resistivity of fresh MSW samples (Shihada *et al.*, 2013), with permission from ASCE

the average physical composition of the landfilled MSW samples, which were compacted to a unit weight of 35 lbs./ft.³ at actual field moisture content in the test box. The resistivity of each sample was plotted against its corresponding paper content and "others" content to find any possible trends. A decrease in the electrical resistivity with increasing paper content and increasing "others" content was observed (Figure 5.18).

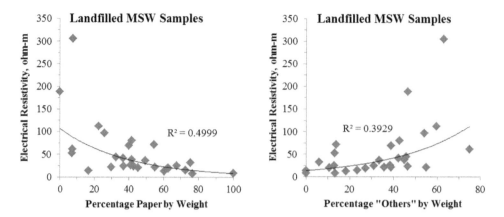

Figure 5.18 Effect of paper content and "others" on electrical resistivity of landfilled samples (Shihada *et al.*, 2013), with permission from ASCE

5.7 Effect of pore fluid composition

Electrical conduction in solid waste is largely electrolytic, taking place in the liquid contained in the pores. In a bioreactor landfill, the liquid contained in the pores can be leachate, re-use water, or a combination of both. The composition of the pore fluid is expected to have a significant effect on the electrical resistivity of MSW.

5.7.1 Using leachate

Leachate usually contains various inorganic ions (ex: $K+$, $Na+$, $NH4+$, $Cl-$). More electrical conduction occurs as a result of the movement of these additional ions in the leachate. A more significant drop in resistivity is expected when MSW is mixed with leachate than when MSW is mixed with water (Yoon and Park, 2001).

   To have a better understanding of the effect of pore fluid composition on the resistivity of MSW, Shihada (2011) measured the electrical resistivity of the five fresh MSW samples. However, leachate collected in October 2010 from an on-site leachate storage tank at the City of Denton Landfill in Denton, Texas, was added to the samples instead of tap water. The fresh MSW samples were first dried completely, then were compacted to a dry unit weight of 35 lb./ft^3 and transferred to the resistivity box. Leachate was added daily to the samples, and the corresponding moisture content was determined. The samples stayed in the test box for at least 24 hours to ensure even moisture distribution, and then the resistivity tests were conducted. Shihada (2011) also conducted the same test using tap water instead of leachate, to compare and investigate the effects of the leachate on the resistivity imaging.

   Figure 5.19 presents a comparison of the resistivity results of the five fresh MSW samples prepared using leachate and those of the samples that were prepared using tap water. The resistivity values of the samples prepared using leachate were slightly lower than the values obtained when the samples were prepared using tap water. A difference of less than 10% was usually observed.

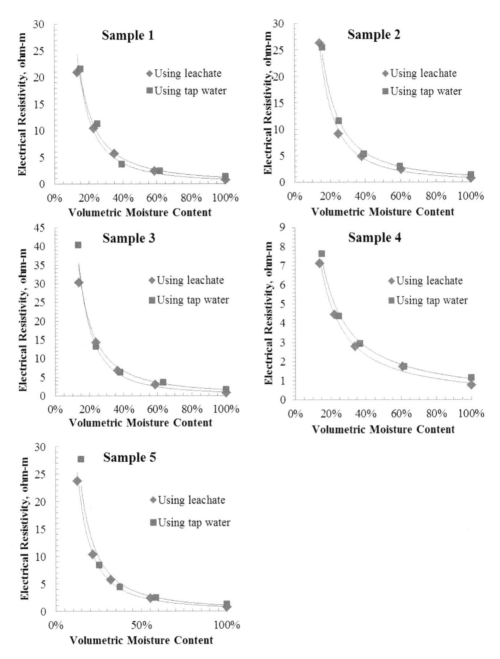

Figure 5.19 Effect of using leachate as the pore fluid on resistivity of MSW (Shihada *et al.*, 2013), with permission from ASCE

Table 5.3 Resistivity and salinity of leachate before and after mixing with MSW (Shihada
et al., 2013), with permission from ASCE

Liquid	Resistivity (Ωm)	Resistivity after mixing with MSW (Ωm)	Salinity (%)	Salinity after mixing with MSW (%)
Tap water	30–32	1.1–1.7	0	0.4
Leachate	3.5–3.7	0.67–0.78	0	0.6

Since, the variations of electrical resistivity at different pore water fluids, such as leachate and tap water, were not significant, Shihada (2011) extended the test to explain the results. During this study, the resistivity of leachate was measured at the end of each test by collecting the liquid and measuring its conductivity with a conductivity meter. Similarly, author used a salinity meter to measure the salinity of the leachate before and after conducting the test. Due to mineralization, the resistivity of the leachate dropped from the 3.5–3.7 ohm-m range to the 0.67–0.78 ohm-m range, as depicted in Table 5.3. A slight increase in salinity from 0% to 0.6% was observed, indicating that salinity is not a factor that should be considered. Similarly, when the resistivity tests were conducted on the same samples using tap water, the resistivity of the tap water decreased from the 30–32 ohm-m range to the 1.1–1.7 ohm-m range. These results indicated that when a liquid is added to a waste sample, the liquid extracts the soluble inorganic and organic compounds present in the waste, and the resistivity of the liquid stabilizes at some constant value after approximately five days. In other words, due to the presence of high amounts of inorganic and organic compounds in the waste itself, the content of the waste is more controlling than the type of fluid in the pores.

5.7.2 Using re-use water

In many bioreactor landfills, when the amount of leachate produced at the landfill is insufficient, the re-use water is recirculated into the waste mass. The re-use water is treated wastewater effluent from an on-site wastewater treatment plant. Shihada (2011) measured the resistivity of the five fresh MSW samples, using re-use water.

During the tests, each of the fresh MSW samples was dried completely, compacted to a dry unit weight of 35 lbs./ft³, and then transferred to the resistivity box. Re-use water was added daily to the sample, and the corresponding moisture content was determined. Shihada left the sample in the test box for at least 24 hours to ensure even moisture distribution and then conducted the resistivity tests and compared the results with those obtained from adding tap water and re-use water.

Figure 5.20 presents a comparison of the resistivity results of the five fresh MSW samples prepared using re-use water and the test results of the samples that were prepared using tap water. Shihada (2011) indicated that the resistivity values of the samples prepared using re-use water were almost identical (less than 10% most of this time) to those obtained when the samples were prepared using tap water.

The resistivity of the re-use water ranged between 13.8 and 14.0 ohm-m initially before it was added to the waste. This resistivity range is between that of tap water and leachate.

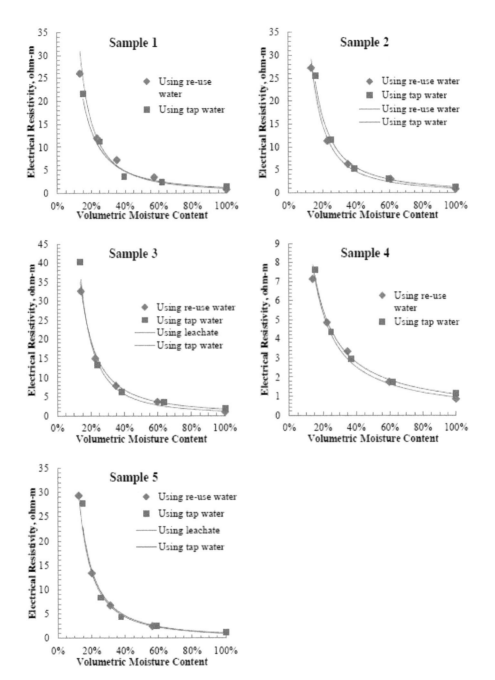

Figure 5.20 Effect of using re-use water as the pore fluid on electrical resistivity of MSW (Shihada, 2011)

Table 5.4 Resistivity and salinity of re-use water before and after mixing with MSW (Shihada, 2011)

Liquid	Resistivity (Ωm)	Resistivity after mixing with MSW (Ωm)	Salinity (%)	Salinity after mixing with MSW (%)
Tap water	30–32	1.1–1.7	0	0.4
Re–use water	13.8–14.0	0.75–1.01	0	0.4

After adding the re-use water to the waste and conducting the resistivity tests, the resistivity of the re-use water dropped to the 0.75–1.01 ohm-m range, as shown in Table 5.4. In addition, a slight increase in salinity, from 0% to 0.4%, was observed, indicating that salinity is not a factor that should be considered.

These results confirm again that due to the presence of high amounts of inorganic and organic compounds in the waste itself, the content of the waste is more controlling than the type of fluid in the pores.

5.8 Statistical model to connect resistivity with MSW properties

Shihada (2011) conducted multiple linear regression (MLR) analyses to develop a model between the electrical resistivity of MSW and the several factors that affect resistivity. Regression analysis is a statistical tool for modeling and analyzing the relationships between several variables, and MLR is used to model the relationship between a response variable and two or more predictor variables. In addition, MLR is valuable tool for simultaneously quantifying the effects of various factors on a single dependent variable. Shihada (2011) utilized a dataset, consisting of 37 observations from the experimental results on different waste samples, to develop the model. The dataset covered the ranges for the predictor variables: moisture content (M/C): 13% – 70%, unit weight: 30 lbs./ft^3–70 lbs./ft^3, percentage paper: 0% – 100%, percentage "others": 0% – 75%, organic content (O/C): 44% – 85%. The response variable was electrical resistivity. With several trial analyses, it was determined that the best model to connect the electrical resistivity with the moisture content, unit weight, paper content, and "other" content is the one presented in Equation 5.3:

$$logR = 3.35056 - 0.0240825W - 0.1936\gamma - 0.018156P + 0.00023668W * P \qquad (5.3)$$

Where,
R = Electrical Resistivity in Ohm-m
W = moisture content in percentage (wet-weight basis)
γ = unit weight in lb./ft^3
P = paper composition in percentage

Shihada's model is a 3-D model that is difficult to imagine. One variable (percentage of paper) was fixed, and a 3-D plot was prepared, using MATLAB. Figure 5.21 presents the surface plot of the model for paper compositions of 20% and 40%.

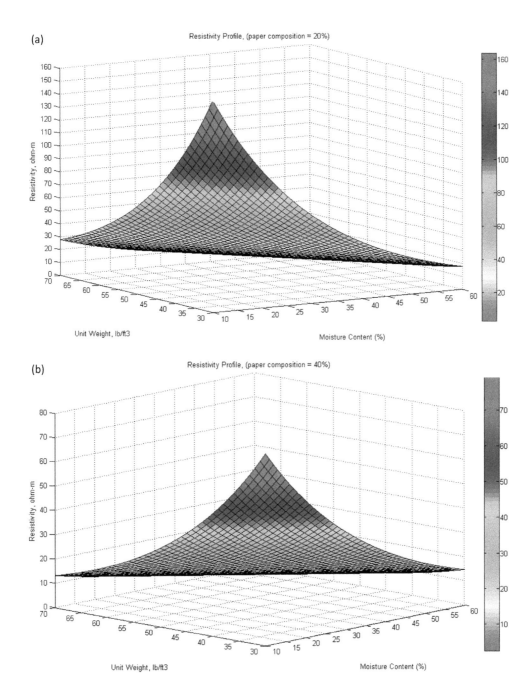

Figure 5.21 Surface plot of Shihada's model for: (a) Paper composition = 20%, and (b) Paper composition = 40% (Shihada, 2011)

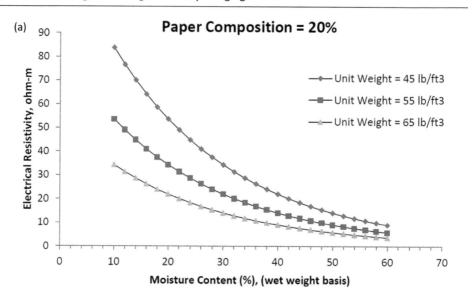

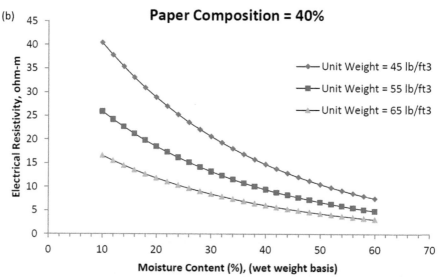

Figure 5.22 Plot of Shihada's model with various unit weights: (a) Paper composition = 20%, and (b) a. Paper composition = 40%. (Shihada, 2011)

The 2D plot (set of curves) with two variables, unit weight (45, 55, and 65 lbs./ft³) and percentage paper (20% and 40%) is presented in Figure 5.22. The unit weight of MSW in a landfill depends primarily on the compaction practices used in the field, the depth of the sample (overburden pressure), the age of the waste, and the composition of the waste. Studies have shown that the unit weight of MSW increases with depth and decomposition. Also, high amounts of soil and fines tend to increase the unit weight of MSW.

5.9 Model validation

Shihada's model was validated based on the 2D electrical resistivity imaging test in the field, the collection and testing of samples from the testing locations, and determining the test parameters in the laboratory. The moisture content was estimated, using the field resistivity values and the developed model. The estimated moisture content was then compared with the laboratory-measured moisture content. Details of the procedure are discussed in the next section.

5.9.1 Resistivity survey in the field

Shihada (2011) conducted 2D electrical resistivity imaging at the landfill between boreholes B70 and B72 in March 2011, at the City of Denton Landfill, in Denton, Texas (Figure 5.23). 56 electrodes, placed at intervals of 6 ft., were utilized in a dipole-dipole array. A programmable eight-channel SuperSting R8/IP resistivity meter was used. The collected data was processed, using Earth Imager 2D software. This software uses a forward modeling subroutine to calculate apparent resistivity values, which are then inverted, using a nonlinear least-squares optimization technique. The resistivity profile between BH 70 and BH 72 is presented in Figure 5.24.

Figure 5.23 Resistivity imaging line between Well 70 and Well 72 (Shihada, 2011)

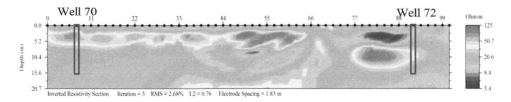

Figure 5.24 Resistivity profile between BH 70 and BH 72 (Shihada, 2011)

5.9.2 Moisture content estimation

Shihada's model was rearranged to calculate the moisture content directly from the electrical resistivity imaging value. The rearranged model is presented in Equation 5.4.

$$W = \frac{3.35056 - logR - 0.1936\gamma - 0.018156P}{0.0240825 - 0.00023668P} \tag{5.4}$$

Where,
R = Electrical Resistivity in Ohm-m
W = moisture content in percentage (wet-weight basis)
γ = unit weight in lb./ft3
P = paper composition in percentage

Before the electrical resistivity values from the field can be used in the model, it has to be corrected to a standard temperature of 70°F. The resistivity survey was conducted on 25 March 2001. On that day, the low daily temperature was 49°F, the high daily temperature was 71°F, and the mean daily temperature was 60°F. These temperatures represent the ambient air temperature and do not necessarily represent the waste temperature at different depths. Temperature within the waste mass is expected to increase with depth. Therefore, the temperature profile within the waste is estimated by following the method used by Yesiller et al., 2005 and Liu, 2007, where the baseline waste temperatures can be estimated using the analytical formulation for sinusoidal fluctuation of temperature with depth (Equation 5.5).

$$T_{(x,t)} = T_m - A_s e^{-x\sqrt{\pi/365s\alpha}} Cos\left[\frac{2\pi}{365s}\left(t - t_0 - \frac{x}{2}\sqrt{\frac{365s}{\pi\alpha}}\right)\right] \tag{5.5}$$

Where,
T(x,t) = temperature (°F) at depth x and time t
Tm = mean annual earth temperature (°F)
As = amplitude of surface temperature wave (°F) x = depth below surface (m)
s = 86,400 seconds
α = thermal diffusivity (m2/day)
t = time of year in days (where 0 = midnight 31 December)
t0 = phase constant = 34.6 days = 2,989,440 seconds

These baseline waste temperatures represent seasonal temperature variations in the waste, but they neglect the internal heat generation that is due to waste decomposition. According to Hanson et al. (2010), the temperature of the waste initially increases with depth, reaching a peak value at middle depth, and then starts to decrease slightly. A study done on waste of different ages (Figure 5.25) showed that the temperature profiles shifted to the left with age, indicating that most of the heat generation occurs during the early years of waste placement.

Since Shihada (2011) conducted the resistivity survey on a closed section of the landfill in which the waste was approximately 20 to 25 years old, it is reasonable to estimate the temperature profile, using equation (5.5), neglecting the heat generated by the waste. A mean annual earth temperature of 67°F was used for the Dallas-Fort Worth area. An amplitude of surface temperature wave of 59°F (15°C) and a thermal diffusivity of 5 x 10–7 m²/s were used, as suggested by Liu, 2007. The resulting temperature profile is presented in Figure 5.26.

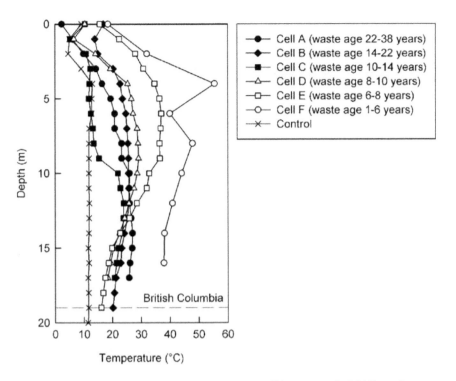

Figure 5.25 Effect of waste age on waste temperature (Hanson et al., 2010), with permission from ASCE

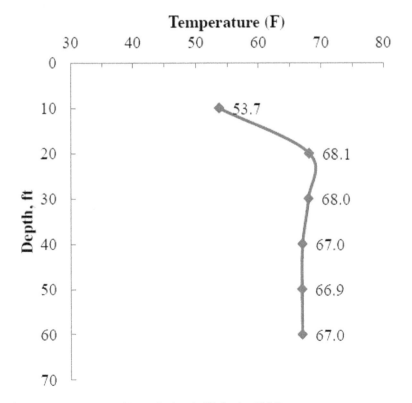

Figure 5.26 Temperature profile with depth (Shihada, 2011)

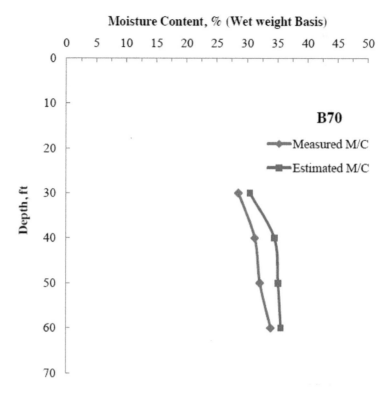

Figure 5.27 Estimated and actual moisture content profile for borehole B70 (Shihada, 2011)

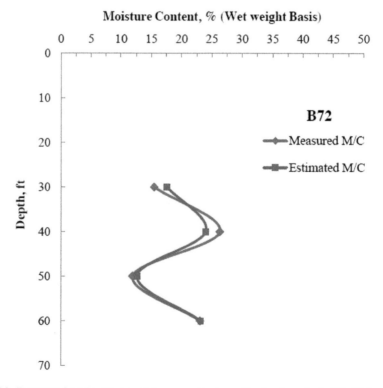

Figure 5.28 Estimated and actual moisture content profile for borehole B7 (Shihada, 2011)

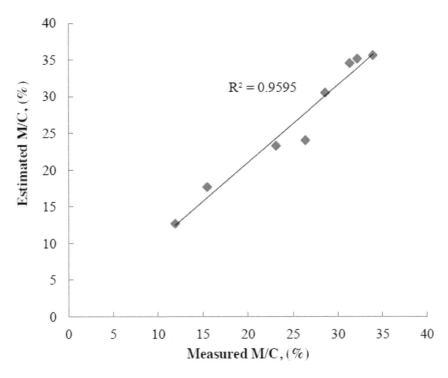

$R^2 = 0.9595$

Figure 5.29 Estimated vs. actual moisture contents for all samples (Shihada, 2011)

Table 5.5 Estimated moisture content for samples from borehole B70 (Shihada, 2011)

Depth (ft.)	Depth (m)	% Paper	Moisture Content (%)	Unit Weight (lb./ft.³)	Resistivity, ohm-m	M/C from Resistivity (%)	% Error
30	9.1	43.9	28.61	45	18.4	30.49	6.6
40	12.2	43.9	31.32	45	16.2	34.53	10.2
50	15.2	43.9	32.16	46	15.2	35.14	9.3
60	18.3	43.9	33.94	48	13.7	35.60	4.9

Table 5.6 Estimated moisture content for samples from borehole B72 (Shihada, 2011)

Depth (ft.)	Depth (m)	% Paper	Moisture Content (%)	Unit Weight (lb./ft³)	Resistivity, ohm-m	M/C from Resistivity (%)	% Error
30	9.1	15.8	15.52	60	34.9	17.66	13.8
40	12.2	15.8	26.39	60	25.9	24.03	−9.0
50	15.2	15.8	11.94	80	18.1	12.64	5.9
60	18.3	15.8	23.15	80	11	23.28	0.5

To estimate the moisture content of MSW using the field resistivity values, average percentages of paper of 43.9% and 15.8% (from physical composition results) were used for samples from B70 and B72, respectively. The unit weight of the samples was estimated according to the compaction practices followed at the landfill. Usually, an average unit weight of 1000 lbs./yd³ (40 lbs./ft³) is typical for fresh waste in the top layers of the landfill. For B70, a unit weight of 40 lbs./ft³ was used for samples from the top layers. Higher unit weights were used for samples at higher depths to account for the effects of overburden pressure. The characterization test of the samples from borehole B72 indicated a high percentage of cover soil and low percentage of paper products. To account for the higher fine content in the waste, a unit weight of 60 lbs./ft³ was used for samples from the top layers in B72.

The moisture contents, estimated using Shihada's model, were compared with the measured moisture contents, and the percentage of error was determined. Good agreement was found between the estimated and measured moisture contents. The results are presented in Tables 5.5 and 5.10 for B70 and B72, respectively. The percentage of error ranged from 4.9% to 10.2% and from 0.5% to 13.8% for samples from B70 and B72, respectively. Figure 5.27 and Figure 5.28 present the estimated and measured moisture content profiles with depth for MSW samples from boreholes B70 and B72, respectively. A plot of the estimated versus the measured moisture contents for all of the samples is presented in Figure 5.29.

Since Shihada's model's estimated moisture content was in good agreement with the actual moisture content, it can be utilized to estimate the moisture content of MSW from an electrical resistivity imaging test with a low margin of error.

Chapter 6

Case studies on the application of resistivity imaging

6.1 General

Resistivity Imaging (RI) has been used in various applications of subsurface investigation, geohazard evaluation, and forensic study. This chapter presents some case studies where RI has been successfully used to evaluate mechanically stabilized earth, earthen dams, and slope stability, and to investigate bridge foundations and bioreactor landfills. However, it should be mentioned that the use of RI is not limited to these field. It can be used as a compliment to conventional testing in most geotechnical, geoenvironmental, and geological studies. Based on the authors' experience, interpretation of RI should be performed with adequate engineering judgements, on a case-by-case basis, and the results should be confirmed by borings and/or other methods. The authors' experience indicates that RI results always provide important information that is necessary for drawing conclusions.

6.2 Application of resistivity imaging in the MSE wall

A forensic study was performed on a MSE wall located at State Highway 342 in Lancaster, Texas, where the top of the wall had experienced 12 to 18 inches (300 to 450 mm) of movement in some sections. The study included soil test borings, installation of an inclinometer, a survey, and resistivity imaging (RI).

A subsurface exploration program was conducted on 15 October 2009. Two test borings, labeled BH-1 and BH-2, were drilled, as presented in Figure 6.1. The drilling was performed using a truck-mounted rig. Soil samples were collected at every 5 ft. throughout the boring. The borehole log is presented in Appendix C (Figure C1 and Figure C2).

The grain-size distribution results and moisture variations are presented in Figure 6.2 and Figure 6.3, respectively. Grain-size analyses of the reinforced fill samples indicated that the percent passing through a No. 200 (0.075 mm) sieve ranged from about 29% to 39%, and the plasticity indices of the sample ranged from 6 to 10. The reinforced soils were classified as clayey sand (SC), according to the Unified Soil Classification System (USCS). As presented in Figure 6.3, the moisture content was higher near the top of the MSE wall. Below 5 ft., the moisture content was almost uniform, and varied within a range of 7% to 10% up to 25 ft. depth. It should be noted that based on the soil boring, moisture content of 7% to 10% is not much, and should not create any problem to the MSE wall.

Since the MSE wall facing indicated bulging in some panels and the select fill contained more than 15% fine content, RI tests were conducted in the reinforced fill area. The layout of the 2D RI line is presented in Figure 6.1. The field setup for RI is presented in Figure 6.4

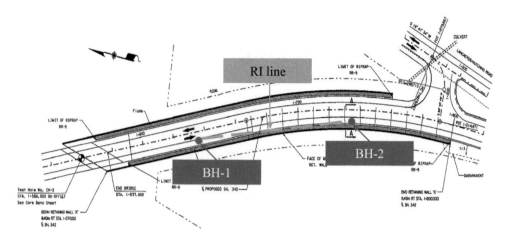

Figure 6.1 Location of soil test boring

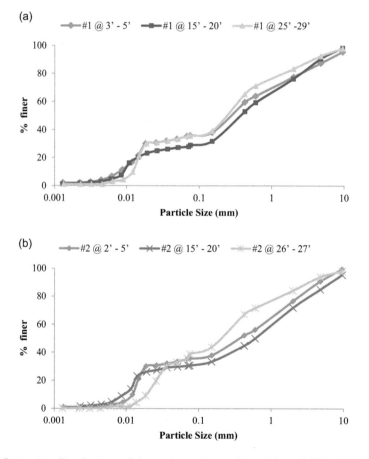

Figure 6.2 Grain size distribution of the soil samples collected from MSE wall: (a) Borehole 1 (b) Borehole 2

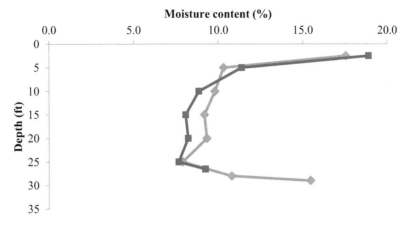

Figure 6.3 Moisture profile of soil specimens

Figure 6.4 Operational setup of resistivity imaging: (a) Drilling of pavement, (b) Insertion of electrode, (c) Connection of cable with equipment, and (d) Layout of cable. (Hossain *et al.*, 2012), with permission from ASCE

The RI test was performed on 5 March 2010. 28 electrodes were spaced 6 ft. apart, and the test covered a length of 168 ft. The RI was conducted using dipole-dipole array. From the test results near borehole 1, perched water or saturated backfill was observed between depths of 12 to 22 ft. (3.6 to 6.7 m). This is labeled as the Perched Water Zone 1 in Figure 6.5. Another perched water zone was found approximately 57 ft. (17.3 m) south of the first inclinometer, between depths of 10 to 20 ft. (3 to 6 m) and is noted as Perched Water Zone 2. A relatively high resistivity zone (Area A, Resistivity > 100 Ωm) was located approximately 7 ft. (2.1 m) below the top of the MSE wall between perched water zones 1 and 2. A comparatively low resistivity zone (Area B, Resistivity < 20 Ωm) was located between the perched water zones at depths of 20 ft. to 30 ft. (6 to 9.1 m), as shown in Figure 6.5.

The perched water zone might have developed due to poor drainage of the backfill. The grain-size distribution curve indicated that the reinforced backfill contained more than 15% fines, which may have significantly reduced the permeability of the soil, resulting in a perched water zone. This perched water had the potential to increase lateral pressure on the wall facing, resulting in the bulging of the facing. From the site investigation, it was observed that the alignment of the facings was not uniform in some locations, as presented in Figure 6.6.

On 22 July 2010, an additional RI test was conducted along the same location to investigate the effects of seasonal variations on perched water locations. The results showed that the perched water zones were increased compared to March 2010. Simultaneously, the high resistance area (Area A) decreased and the low resistance area (Area B) increased, both significantly, as presented in Figure 6.7.

The initial site investigation of the MSE wall, using soil borings, indicated that the moisture content was lower (7–10%) along the depth of the borehole, especially at 15 ft. depth, where the bulging was located. The continuous profile of the subsurface of the soil

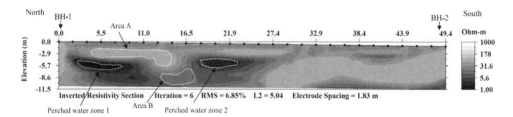

Figure 6.5 Resistivity image conducted in March 2010 (Hossain *et al.*, 2012), with permission from ASCE

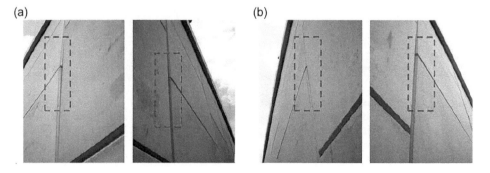

Figure 6.6 Bulging of MSE wall

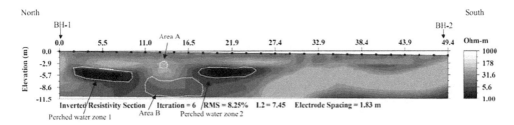

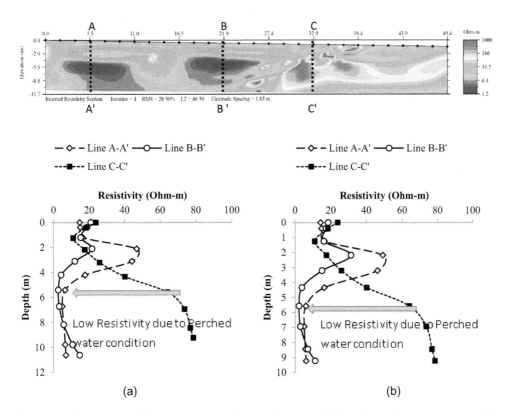

Figure 6.7 Resistivity image conducted in July 2010 (Hossain *et al.*, 2012), with permission from ASCE

Figure 6.8 Variations of resistivity along the depth of the MSE wall at different sections: (a) RI conducted on March 2010, and (b) RI conducted on July 2010

was investigated, using 2D RI. The location and extent of the low resistivity area was determined and indicated the existence of the perched water zone. As presented in Figure 6.8, the resistivity of the reinforced soil was significantly low due the presence of the high moisture along Line A-A and Line B-B, compared to Line C-C. The vertical profile of resistivity was extracted from the 2D profile to better understand the results. The study clearly indicated the usefulness of the RI techniques for investigating the moist zone in the reinforced backfill of the MSE wall.

6.3 Application of RI in the investigation of slope failure

Each year, rain-induced slope failures cause significant damage to highway infrastructures and environments all around the world. Rainfall-induced slope failure is a common problem in the North Texas area, where the slopes are constructed on high plasticity clays. With the improvements in the geophysical data acquisition system, the 2D RI test aids in understanding the *in-situ* behavior of soil due to rainwater infiltration, as well as associated changes in shear strength. During the infiltration process, the matric suction in an unsaturated soil slope decreases as the saturation increases with time, thereby reducing the shear strength of the soil. To investigate a rainfall-induced highway slope failure, it is important to identify the depth of the moisture variation zone (i.e., active zone) where the changes in matric suction usually take place. The 2D RI test was conducted to investigate the slope failures of three highway slopes in the North Texas area, and the results are presented here.

6.3.1 Failure investigation of a highway slope over highway US 287

The study was conducted on a slope along highway US 287 S., near the St. Paul overpass in Midlothian, Texas. The slope is about 30 to 35 ft. (9.2 to 10.6 m) high, with an angle of 3H: 1V. The surface of the slope is covered with naturally occurring grass. Tension cracks were identified in the shoulder of the slope, as presented in Figure 6.9. The cracks had developed due to the surficial movement of the slope and the low confining pressure at the surface that allows clay to crack at relatively low matric suction. These cracks generally become closed with depth, as the confining pressure increases (Khan *et al.*, 2017).

A full-scale field investigation was conducted to investigate the reasons behind the surficial movement of the slope. Three soil test borings were performed at the crest of the slope, as presented in Figure 6.9. Soil samples were collected from different depths to determine the geotechnical properties of soils. In addition, Texas cone penetrometer (TCP) tests were conducted at borehole locations. The predominant strata below the slope surface was a dark

Figure 6.9 Tension cracks along the shoulder (Khan *et al.*, 2017), with permission from ASCE

brown high plasticity clay with calcareous nodules. The subsurface conditions were also inter-layered with yellow brown clay. A thin layer of crushed limestone was observed within the top 3.2–5 ft. (1–1.5 m) of the slope at several locations. The upper part of the slope was rich in cracks and fissures. The maximum width of the open cracks was about 2 in (50 mm) at the crest of the slope. As a part of the investigation, a subsurface exploration program was conducted in October 2010. The layout of boreholes is presented in Figure 6.10.

Three soil test borings were performed near the crest of the slope, with the depths ranging from 20 ft. – 25 ft. Both the disturbed and undisturbed soil samples were collected from different depths and were tested to determine the geotechnical properties of the subsoil. Based on the results of the laboratory investigation, all of the collected soil samples were classified as high plastic clay (CH) soil, according to the Unified Soil Classification System (USCS). The liquid limits and the plasticity indices of the samples ranged between 48–79 and 25–51, respectively. The moisture profiles on the depth and plasticity chart along the three bore holes are presented in Figure 6.11. In addition, the borehole logs are attached in Appendix C (Figure C3 to Figure C5). The moisture profile indicated an increase in moisture below 5 ft. that ranged up to 20 ft.

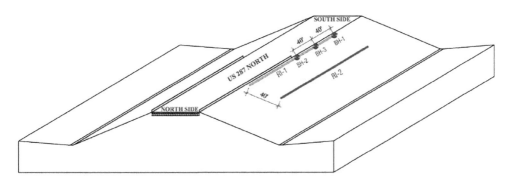

Figure 6.10 Layout of boreholes and resistivity imaging lines

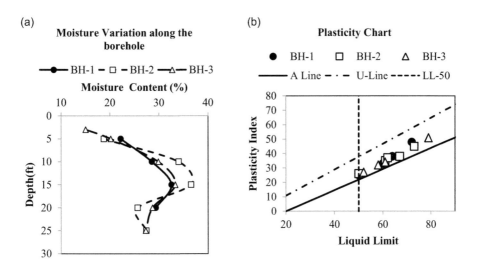

Figure 6.11 Laboratory test results: (a) Moisture distribution along the boreholes (b) Plasticity chart along the boreholes (Khan et al., 2015), with permission from ASCE

The project site is located in the Eagle Ford geological formation, which is composed of residual soils consisting of clay and weathered shale (shaley clay), underlain by unweathered shale (USGS, 2013). The weathered shale contains gypsum in-fills and debris and is jointed and fractured with iron pyrites. The unweathered shale is typically gray to dark gray and commonly includes shell debris, silty fine sand particles, bentonite, and pyrite. The Eagle Ford formation consists of sedimentary rock that is in the process of degrading into a soil mass. This formation also contains smectite clay minerals and sulfates. It should be noted that smectite clay minerals are highly expansive in nature.

With the advancement of new software for the interpretation of resistivity measurements, 2D resistivity imaging (RI) is extensively used in shallow geophysical investigations and geo-hazard studies (Hossain *et al.*, 2010). During the current study, the RI test was used to inves-tigate the subsurface condition of the US 287 slope. Two 2D RI lines, designated as RI-1 and RI-2, were conducted at the slope. RI-1 was conducted at the top of the slope near the crest, as presented in Figure 6.12. RI-2 was conducted at the middle of the slope, 40 ft. from RI-1.

The RI investigations were conducted using eight-channel Super Sting equipment, which is faster than the conventional single-channel unit. A total of 56 electrodes were utilized dur-ing the resistivity imaging. The length of the investigated line was 275 ft., with electrode spacing of 5 ft. c/c. The 2D RI profiles along RI-1 and RI-2 are presented in Figure 6.13 (a) and Figure 6.13 (b), respectively. In addition, the variations of resistivity along the boreholes are presented in Figure 6.14.

Based on the 2D RI profile, a low resistivity zone was observed near the top soil at both RI-1 (at crest) and RI-2 (middle of the slope). The resistivity of slope significantly decreased up to 16.4 ohm-ft. at depths from 5 ft. to 14 ft. It should be noted that the significant low resistivity might have occurred due to the presence of high moisture in the soil.

The RI profile along borehole BH-2 in Figure 6.14 was utilized to predict the mois-ture content of the soil, using the model proposed by Kibria and Hossain, 2016. During this analysis, the average liquid limit of the soil was considered as 70, with a void ratio of 0.8. The predicted moisture content from RI and the moisture content from the site investigation are presented in Figure 6.15. The predicted moisture content fitted very well with the actual moisture content along borehole 2 of the US 287 slope.

Figure 6.12 Resistivity imaging field setup of: (a) RI-1 and (b) RI-2 (Khan et al., 2015), with permission from ASCE.

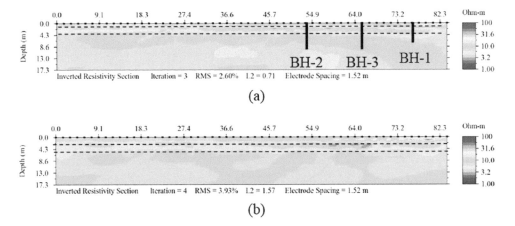

(a)

(b)

Figure 6.13 Resistivity imaging at the US 287 slope: (a) Resistivity profile for RI-1, (b) Resistivity profile for RI-2, and (c) Variations of resistivity along the boreholes (Khan et al., 2015), with permission from ASCE

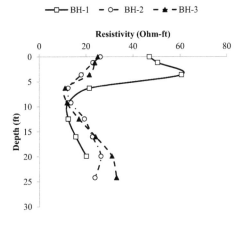

—□— BH-1   – ○– BH-2   – ▲ – BH-3

Figure 6.14 Variations of resistivity along the boreholes (Khan et al., 2015), with permission from ASCE

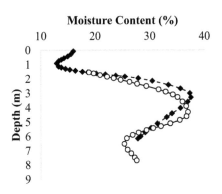

– ◆ – From RI value —○— From Borehole

Figure 6.15 Comparison of moisture content for actual vs. model-predicated values

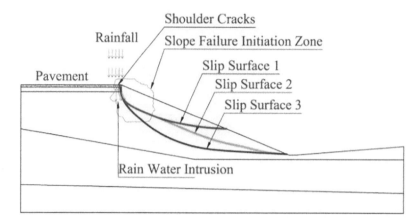

Figure 6.16 Schematic of Possible Surficial Slope Failure (Khan *et al.*, 2015), with permission from ASCE

The subsoil investigation results indicated that the US 287 slope was constructed using high plastic clay; the dominant mineral of the soil is montmorillonite. The high plastic clay, with the presence of montmorillonite, makes it highly susceptible to swelling and shrinking upon wetting and drying. It should be noted that fully softened strengths are eventually developed in high plastic clays in field condition after being exposed to environmental conditions (i.e., shrink and swell, wetting-drying, etc.) and provide the governing strength for first-time slides in both excavated and fill slopes (Saleh and Wright, 1997). The reduction in friction angle is not significantly due to cyclic wetting and drying of soil; however, the cohesion of the soil almost disappears in the fully softened state (Saleh and Wright, 1997). The near-surface soil at the US 287 slope may have been softened due to shrinkage and swell behavior which led to the initiation of the movement of the slope and resulted in the crack over the shoulder.

Based on the subsoil investigation and resistivity imaging, it was evident that a high moisture zone existed between 5 ft. and 14 ft., near the crest of the slope. The shoulder crack provided easy passage of rainwater into the slope, which eventually led to saturation of the soil near the crest. As a result, the driving forces increased, which decreased the factor of safety. It should be noted that the US 287 slope did not fail during the investigation. Based on the historical information, the slope had the potential to fail within the next few years, as the movement had initiated at the crest, which is an indication of the beginning of failure. The initiated movement might follow any of the possible slip surfaces, as presented in Figure 6.16.

6.3.2 Failure investigation of a highway slope over highway SH 183

The investigated slope was located along SH 183, east of the exit ramp from eastbound SH 183 to northbound SH 360, in the northeast corner of Tarrant County. The schematic and failure photos of the slope are presented in Figure 6.17 and Figure 6.18, respectively. The height of the slope was approximately 10.97 m (36 ft.), with a slope geometry of 2.5(H): 1(V).

The project site was located on the Eagle Ford formation and contained shale, siltstone, and limestone. In the upper part, the formation was lime and shale, yellowish-brown, and

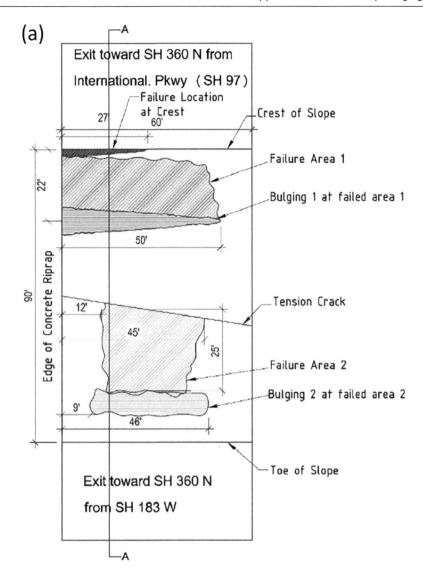

Figure 6.17 Failure condition of the SH 183 slope: (a) Front elevation view, (b) Section A-A (Khan *et al.*, 2017)[1]

flaggy. The lower part of the formation was composed of silt and very fine-grained sandstone, yellow to gray in color, and mostly laminated flaggy; however, it should be noted that some limestone is silty, medium-brown in color, and laminated (USGS, 2013). The Eagle Ford formation consisted of sedimentary rock that was in the process of degrading into a soil mass. This formation also contained smectite clay minerals and sulfates. The swell potential,

1 Reprinted from Engineering Geology, Vol 219, Mohammad Sadik Khan, Sahadat Hossain, Asif Ahmed, Mohammad Faysal, Investigation of a shallow slope failure on expansive clay in Texas, 118-129, Copyright (2017), with permission from Elsevier.

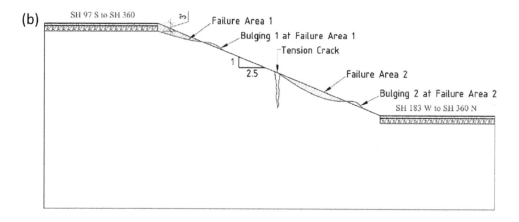

Figure 6.17 Continued

Figure 6.18 Photo of slope failure: (a) Crack on shoulder (b) Failed slope near crest (c) Bulging (d) Tension crack (Khan *et al.*, 2017)[2]

2 Reprinted from Engineering Geology, Vol 219, Mohammad Sadik Khan, Sahadat Hossain, Asif Ahmed, Moham-mad Faysal, Investigation of a shallow slope failure on expansive clay in Texas, 118-129, Copyright (2017), with permission from Elsevier

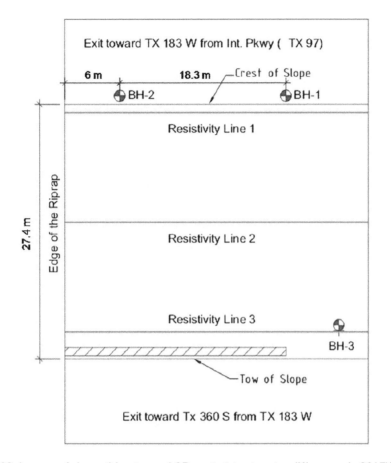

Figure 6.19 Layout of the soil boring and 2D resistivity imaging (Khan *et al.*, 2017)[3]

compressibility, and creep deformation were expected to be high in the Eagle Ford Shale due to the high percentage of smectite (Hsu and Nelson, 2002). This could lead to problems such as slope failures, foundation damage, mine failures, and shale embankment failures (Abrams and Wright, 1972).

The site investigation included a collection of samples from three test borings, designated as BH-1, BH-2, and BH-3. BH-1 and BH-2 were located near the crest of the slope, and BH-3 was located at the toe of the slope. The collected samples were tested in the laboratory to investigate the soil index properties and shear strength parameters that would be used in further analyses. Geophysical investigation was carried out using 2D electrical resistivity tomography (ERT). The layout of the soil test borings and ERT lines are presented in Figure 6.19.

The soil index properties, moisture content (MC), liquid limit (LL), and the plasticity index (PI) of the soil samples obtained from the site are presented in Figure 6.20. The liquid

3 Reprinted from Engineering Geology, Vol 219, Mohammad Sadik Khan, Sahadat Hossain, Asif Ahmed, Mohammad Faysal, Investigation of a shallow slope failure on expansive clay in Texas, 118-129, Copyright (2017), with permission from Elsevier.

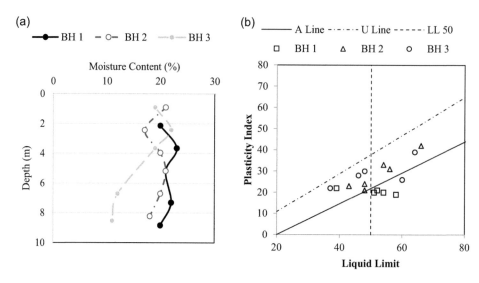

Figure 6.20 Laboratory test results: (a) moisture variation with depth of the boreholes (b) plasticity chart (Khan *et al.*, 2017)[4]

limit and plasticity index tests were conducted in accordance with the ASTM D4318 guidelines. The LL of the soil samples were within a range of 36 to 48 for the top 8 ft., near the crest of the slope for BH-1 and BH-2. The PI values ranged between 20 and 24. The LL ranged between 60 and 64, and PI ranged between 24 and 35 in the top 8 ft. near BH-3. The LL and PI values ranged from 48–64 and 31–42, respectively, for the soil below 8 ft. The soil test results were analyzed according to the Unified Soil Classification System (USCS) and were classified as low-to-highly plastic clay (CL-CH) for boreholes BH-1 and BH-2, and highly plastic clay (CH) for BH-3. The variation of moisture content with the depth of each borehole is shown in Figure 6.20, which indicates that the moisture content of borehole BH-1 was 20% to 23%, while borehole BH-2 had a moisture content of 17% to 21%.

A geophysical study using RI at the slope was conducted using dipole-dipole array, which is widely used for high sensitivity to horizontal changes and better horizontal data coverage for 2D surveys. In this study, three 2D RI tests, designated as RI-1, RI-2, and RI-3, were conducted near the crest, middle, and toe of the slope. Investigations of the subsurface conditions included analysis of the lateral and vertical variations in subsurface moisture content and near-surface geology. The RI tests on the slope were conducted at the end of April 2014, during the early summer period in Texas. Eight-channel Super Sting equipment was employed, with 56 electrodes placed at 1.5 m (5 ft.) c/c intervals, resulting in an 84 m (275 ft.) long ERT line. The recorded data from the RI tests was analyzed using the Earth Imager 2D software, and the obtained resistivity profiles are presented in Figure 6.21.

The resistivity imaging profile revealed a comparatively high resistivity zone, up to the top 2.13 m (7 ft.) depth (Figure 6.21 a) of RI-1, which might suggest the existence of a low

4 Reprinted from Engineering Geology, Vol 219, Mohammad Sadik Khan, Sahadat Hossain, Asif Ahmed, Mohammad Faysal, Investigation of a shallow slope failure on expansive clay in Texas, 118-129, Copyright (2017), with permission from Elsevier.

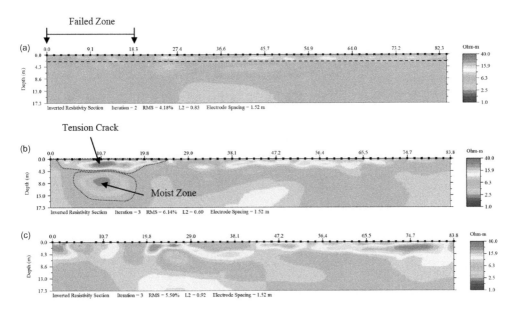

Figure 6.21 (a) Test results of the resistivity Line RI-1(at crest), (b) Test results of the resistivity Line RI-2 (at the middle of the slope), and (c) Test results of the resistivity Line 3 (at toe of the slope) (Khan *et al.*, 2017)[5]

moisture zone near the top of the slope. It should be noted that the ERT test was conducted during the early summer period; therefore, the low moisture zone likely indicated the active zone. A low resistivity zone was also found in RI-1 after 2.13 m (7 ft.) depth, which might indicate the presence of high moisture below the active zone.

Testing at RI-2 was conducted over the tension crack at the middle of the slope (Figure 6.64). A high resistivity zone was observed immediately below the tension crack zone in the resistivity profile of RI-2, where the depth of the high resistivity zone was 3.66 m (12 ft.). Electricity was unable to pass through the tension crack, resulting in very high resistivity. A low resistivity zone was observed below the tension crack, marking the presence of a saturated zone due to rainwater intrusion through the crack. The electrical resistivity tomography Line RI-3 was located at the bottom of the slope; however, no distinct moisture or high resistivity zone was observed.

Another RI test was conducted in July 2016 to investigate the resistivity variations along RI-1 at the failed area near the crest of the slope and to verify the depth of the moisture variation zone. The variations of the ERT at the failure location, between April 2014 and July 2016, are presented in Figure 6.22 (a). Moreover, the variations of resistivity with depth along lines Line-1 and Line-2 are presented in Figure 6.22 (b) and Figure 6.22 (c). Due to the temperature and moisture distribution, evaporation occurring in the surficial soil may have caused moisture variation. For highly plastic clay soil, the depth of the moisture variation

5 Reprinted from Engineering Geology, Vol 219, Mohammad Sadik Khan, Sahadat Hossain, Asif Ahmed, Mohammad Faysal, Investigation of a shallow slope failure on expansive clay in Texas, 118-129, Copyright (2017), with permission from Elsevier.

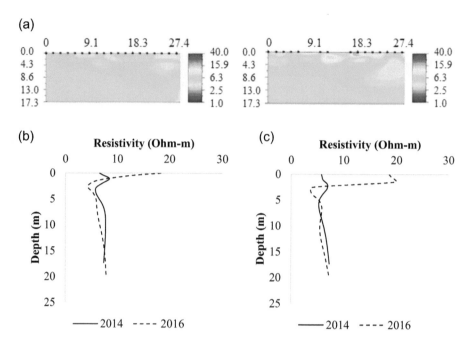

Figure 6.22 (a) Comparison of ERT at failed area during April 2014 and July 2016, (b) Vertical ERT variation along Line 1, and (c) Vertical ERT variation along Line 2 (Khan et al., 2017)[6]

zone is known as the depth of the active zone. The vertical resistivity profiles in Figure 6.22 (b) and Figure 6.22 (c) present the resistivity variations at the upper 2.13 meter (7 ft.) depth at different time periods. The observed resistivity variations occurred as a result of seasonal moisture variation. Therefore, the depth of the active zone was determined to be 7 ft.

The RI profile along Line 1 in Figure 6.22 (a) was utilized to predict the moisture content of the soil, using the model proposed by Kibria and Hossain, 2016. During this analysis, the average liquid limit of the soil was considered as 60, with a void ratio of 0.55, based on the soil test data. The predicted moisture variation from the RI tests and the moisture content from site investigations along BH-2 near the crest of the slope are presented in Figure 6.23. The predicted moisture content from the RI test conducted in April 2014 fitted very well with the actual moisture content along borehole 2 of the SH 183 slope. It should be noted that the site investigation and RI tests were conducted concurrently. The predicted moisture content, using the RI test conducted on July 2016, indicated that the moisture content reduced to 10% near the surface. It should be noted that during the summer, the temperature of the DFW area is often more than 100°F, which dries the highly plastic clayey soil. In addition, shrinkage cracks are commonly observed in the high PI clay. Thus, the predicted moisture content using Kibria and Hossain's, 2016 model fit well with the field condition.

6 Reprinted from Engineering Geology, Vol 219, Mohammad Sadik Khan, Sahadat Hossain, Asif Ahmed, Mohammad Faysal, Investigation of a shallow slope failure on expansive clay in Texas, 118-129, Copyright (2017), with permission from Elsevier.

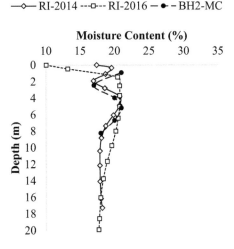

Figure 6.23 Seasonal moisture variations derived from RI test results.

6.3.3 Failure investigation of a highway slope over interstate 30

Investigation of a failed highway slope along the west bound Interstate Highway (IH) 30 near Dallas, Texas was conducted using 2D electrical resistivity imaging. The failure of the slope occurred near the crest, following rainfall. The condition of the slope during the investigation is presented in Figure 6.24.

Ten (10) 2D RI sections were set up at the site. The RI included three (3) 2D lines at the slope failure location, three (3) 2D lines on the west side of the failure location, and four (4) 2D lines on the east side of the failure location. The locations of the RI are presented in Figure 6.25.

Super Sting R8 IP equipment was utilized to conduct RI tests at the site. Tests were performed using 56 electrodes with a spacing of 1.5 m (5 ft.). A dipole-dipole array was selected because a high lateral resolution in a relatively shallow depth was necessary for the assessment of the subsurface condition. Earth Imager 2D software was used to analyze the data obtained from the field. The operational setup of the RI tests is presented in Figure 6.26.

The RI tests were performed on October 2009, January 2010, and April 2010 to determine the variations of moisture contents in different seasons. RI, at the failed sections, is presented in Figure 6.27. A low resistivity area was observed at approximately 2.1 m (7 ft.) depth. The resistivity of the failed section was as low as 1 Ohm-m, indicating the presence of high moisture content at that section during failure.

Line 2 of the resistivity survey was conducted at the middle of the slope, parallel to Line 1. A low resistivity area was at approximately 2.1 m (7 ft.) depth, as presented in Figure 6.27 (b). The lowest resistivity along Line 2 was 1.7 Ohm-m, which was slightly higher than the lowest resistivity along Line 1. Line 3 of the resistivity survey was performed at the bottom of the slope, parallel to Lines 1 and 2. According to the RI results along Line 3, a low resistivity area was identified at approximately 1.5 m (5 ft.) of depth. The lowest resistivity along Line 3 was 2.4 Ohm-m, which was higher than the lowest resistivity of Line 1 and Line 2.

The low resistivity (1 Ohm-m) along Line 1 indicated that the soil contained higher moisture contents at the top of the slope. The RI results indicated that the failure might initiate at the crest, and the depth of the failure might reduce over the length of the slope. It should

Figure 6.24 Cracks and settlements in the failed slope

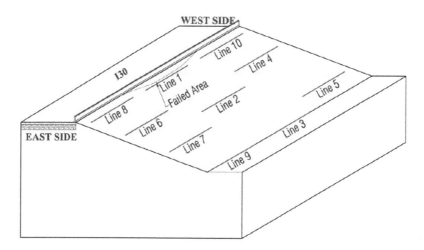

Figure 6.25 Resistivity imaging (RI) sections at the site

Figure 6.26 Operational setup of resistivity imaging (RI)

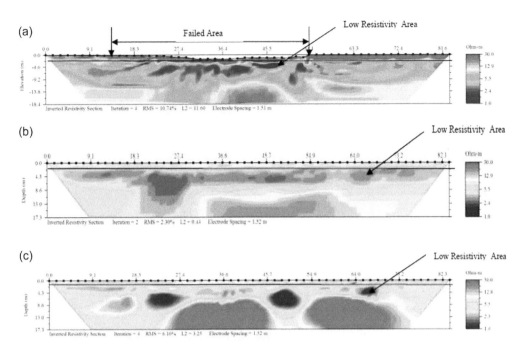

Figure 6.27 Resistivity imaging conducted on October 2009: (a) Line 1, (b) Line 2, and (c) Line 3

be mentioned that the soil from the failed section of the slope was removed after the tests were performed in October 2009; therefore, further investigation was not possible at the failed section.

The vertical RI profile at the failure area of Line 1, Line 2, and Line 3 was utilized to predict the moisture content of the soil, using the model proposed by Kibria and Hossain, 2016. During this analysis, the average liquid limit of the soil was considered as 70 with a void ratio of 0.55, based on the soil test data. The predicted moisture variations from the RI tests are presented in Figure 6.28. Based on the predicted moisture content, the maximum moisture content was observed at the failure area in Line 1, at 3.7 m depth. It should be noted that the maximum moisture content of 35% was observed along the failure plane near the crest of the slope. The predicted moisture content along Line 2 and Line 3 was lower than the moisture content along Line 1. This confirmed the intrusion of rainwater at the crest of the slope, which created a perched water zone. As the slope failed, the perched water was drained away, which reduced the moisture content at the middle and toe of the slope.

The resistivity survey along Line 6 was conducted 6.1 m (20 ft.) from the top of the slope. The RI results of three different months are presented in Figure 6.29. A low resistivity zone was observed at approximately 3 m (10 ft.) depth. The lowest resistivity along Line 6 was 3.7 Ohm-m. However, along the failure area at Line 1, the lowest resistivity was 1.0 Ohm-m at a depth of 2.1 m (7 ft.). Figure 6.29 (b) indicates a low resistivity area at approximately 4.3 m (14 ft.) depth. The observed lowest resistivity along Line 6 was 2.3 Ohm-m. The resistivity of the soil at the top 2.1–2.4 m (7 to 8 ft.) might have increased due to drying of the top soil. A resistivity survey along Line 6 in April 2010 showed that the low resistivity area was approximately at 4.9 m (16 ft.) depth (Figure 6.29 (c)).

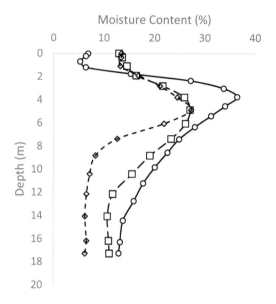

Figure 6.28 Variations of moisture content at the middle of the failure area at the I30 slope

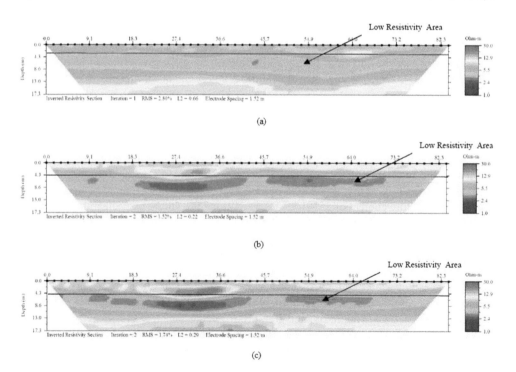

Figure 6.29 Resistivity imaging along Line 6: (a) October 2009, (b) January 2010, and (c) April 2010

The resistivity survey along Line 7 was conducted at 12.2 m (40 ft.) from the top of the slope and parallel to Line 6. The investigation results are presented in Figure 6.30. A low resistivity area was identified at a depth of 2.1 m (7 ft.) during the RI test performed in October 2009. However, a resistivity survey along Line 7 on January 2010 showed that the low resistivity area started from a depth of approximately 2.7 m (9 ft.). The lowest resistance observed along Line 7 was 2.3 Ohm-m. The resistivity value for the top 1.8–2.1 m (6–7 ft.) of soil might have increased due to drying. Furthermore, the resistivity survey along Line 7, conducted on April 2010, suggested that the low resistivity area was located at approximately 3 m (10 ft.) depth. Resistivity results on the west side of the failed area showed trends similar to that of the east side of the failed area.

The depths of high moisture zones were identified from resistivity profiles for the three investigative periods, as presented in Figure 6.31. A linear interpolation method was utilized

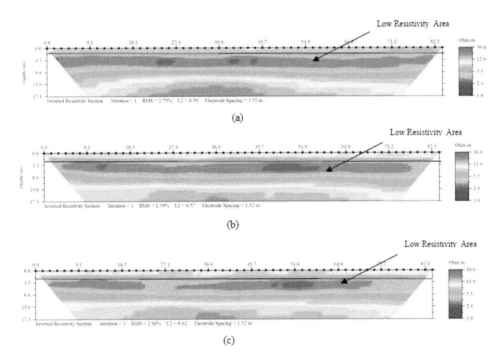

Figure 6.30 Resistivity imaging along Line 7: (a) October 2009, (b) January 2010, and (c) April 2010

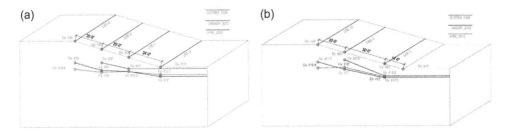

Figure 6.31 Seasonal moisture variations: (a) East side (b) West side of the failed section

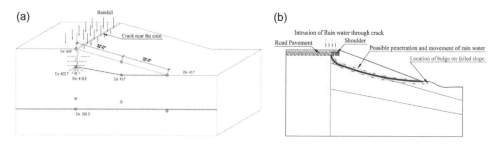

Figure 6.32 Possible mechanism of failure: (a) Intrusion of moisture (b) Possible slip plane

to extend low resistivity zones along the length of the slope. Figure 6.32 also indicates the seasonal variations in moisture and intrusion of water in the investigated slopes. During October 2009, a high moisture zone was located at shallow depths; however, the depth of the moisture profile increased in January 2010.

During the site investigation, cracks and significant settlements, which could have provided easy passage for water, were observed along the shoulder. The intrusion of water eventually caused saturation of soils near the top of slopes and subsequent reduction in shear strength. In addition, an increase of the driving force might have occurred due to pore water pressure. After a few cycles of drying and wetting, the soil strength would have been significantly reduced, which could have caused the failure of a slope. The possible slip surface was determined based on the field observations and RI tests (Figure 6.32).

6.4 Application of RI in the investigation of an Earth dam

Many earth dams and levees have seepage problems which require periodic maintenance to prevent potential catastrophic failure. It is somewhat difficult to identify the exact location of the seepage; however, detection of the seepage path is crucial to maintaining an adequate safety factor of the earth structures. RI technics has been successfully utilized to investigate the seepage path of dams. Two case studies on the investigation of the seepages in dams are presented in the following section.

6.4.1 Investigation of Gladewater Dam

Hubbard (2010) performed a study, using RI, on the Gladewater Dam which is located at Lake Gladewater, Texas. The dam was constructed in 1952, across the waterway of Glade Creek. The height of the dam is approximately 11.6 m (38 ft.) from the toe. The elevations of the dam crest and normal water level of the lake are reported as 95.7 m (314 ft.) and 91.8 m (301 ft.) above mean sea level, respectively. Therefore, a free board of 3.9 m (13 ft.) was allowed at normal water level condition. The total length of the dam is approximately 335 to 366 m (1,100 to 1,200 ft.). The angles of the side slopes range from 1 to 1 (horizontal to vertical) to 2 to 1 (Hubbard, 2010).

A site visit was performed on the Gladewater Dam as a part of a regular monitoring program. The visual inspection indicated the presence of standing water, heavy vegetative growth, localized sloughing, and sliding failures along the downstream slope of the dam.

Several recommendations were provided after the visual assessments, such as further monitoring of the downstream seepage condition, regular maintenance, and repair of a few areas near the toe and face of the downstream slope (TCEQ, 2005). Therefore, an investigation was conducted to identify the seepage path and sloughing of the dam, using soil test borings, RI, and a multichannel analysis of surface waves (MASW) survey.

During field inspection, heavy vegetative growth was observed on the downstream slope near the location of boring BH-3. The investigation results indicated that the moisture contents of subsoils were high along this borehole. In addition, shallow slope failures were identified near borings BH-1 and BH-2. The average SPT blow counts were less than 10 inches between 3.1–10.7 m (10 to 35 ft.) depth along BH-1 and BH-2. The observed standing water and softened soil conditions might have been related to the presence of low-strength materials at different depths. The variations of the moisture content and SPT N values along the depth of the boreholes are presented in Figure 6.33. Soil test borings indicated that the subsurface consisted of clayey sand, sandy lean clay, and lean clay with sand. The ranges of PI and percentages of the samples passing through a No. 200 (0.075 mm) sieve were 9 to 28 and 41% to 76%, respectively. The soil test boring results were in good agreement with the field observation.

In addition to the soil test boring, RI was carried out along the crest of the Gladewater Dam. A total of 28 electrodes with a spacing of 3.1 m (10 ft.) were utilized (length of profile 82.3 m). The approximate depth of investigation was 16.5 m (54 ft.) at this condition (20% of the array length). A dipole-dipole array was selected because high lateral resolution in a relatively shallow depth was necessary for the assessment of the subsurface condition of the dam (Loke, 1999; Hosssain et al., 2010).

A roll-along survey was performed to increase the total length of the investigation. A total length of 335 m (1100 ft.) was investigated, using this method. The RI profile and operational setup at the crest of the dam are presented in Figure 6.34.

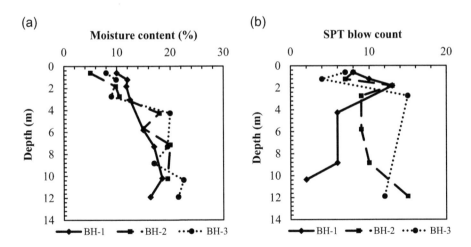

(a) Moisture content

(b) SPT blow count

Figure 6.33 Variation of moisture content and SPT along borehole locations at the Gladewater Dam (Hubbard, 2010)

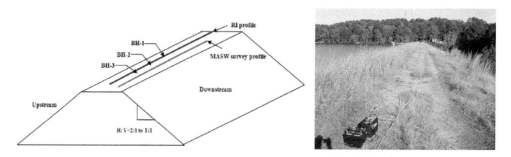

Figure 6.34 Field setup of resistivity imaging (RI) (Hubbard, 2010)

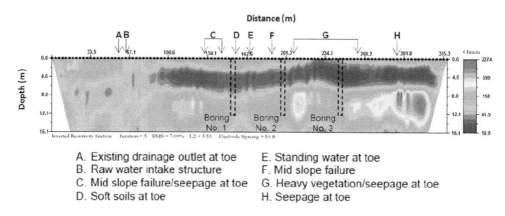

A. Existing drainage outlet at toe
B. Raw water intake structure
C. Mid slope failure/seepage at toe
D. Soft soils at toe

E. Standing water at toe
F. Mid slope failure
G. Heavy vegetation/seepage at toe
H. Seepage at toe

Figure 6.35 Resistivity imaging performed on the dam (Hubbard, 2010)

The 2D RI profile of the test results are presented in Figure 6.35. According to the results, the resistivity of the investigated area ranged from approximately 11 to 2300 Ohm-m. It should be noted that the ranges of measurements corresponded to clay, sands, and gravels (Samouëlian *et al.*, 2005). The lower resistivity values (10.9 to 100 Ohm-m) might be associated with highly saturated clayey sand and sandy clay materials. The resistivity values between 100 and 400 Ohm-m might be representative of the clayey sands and stiffer clay materials. Furthermore, resistivity measurements above 400 Ohm-meter might indicate higher concentrations of sand and gravel.

Figure 6.35 suggests that the resistivity was low between 122 and 183 m (400 to 600 ft.) length. Soil softening, seepage, and shallow slope failure were observed in this location. High vegetation growth was identified near borehole BH-3. The presence of high moisture between 201.2 and 268.3 m (660 to 880 ft.) might cause a reduction in resistivity and enhance vegetation growth in this area.

The variations of resistivity along borehole locations are presented in Figure 6.36. It should be noted that a significant increase in moisture was identified at 3.9 m (13 ft.) depths in boreholes 2 and 3 (Figure 6.36), which might have caused a reduction in resistivity at a depth of 3.9 m (13 ft.). The moisture variations from the field testing results also indicated

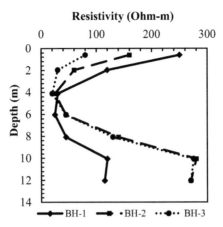

Figure 6.36 Variations of resistivity along the depth of the boreholes in Gladewater Dam (Hubbard, 2010)

a high moisture content from 3.9 m (13 ft.) depth. Thus, the 2D RI test results were in good agreement with the conventional soil boring.

It is evident that the low resistivity zone indicated the seepage path of the earth dam. Thus, the exact extent and depth of the possible seepage area can be detected by looking at the 2D RI profile in Figure 6.35.

6.4.2 Investigation of the Lewisville Dam

The Lewisville Dam was constructed in 1954 in Denton County, near Lewisville, Texas, on the Elm Fork of the Trinity River (Figures 1.1 and 1.2). It was built with impermeable earthen materials on top of the bedrock, impervious stiff clay, semi-impervious sandy clay, and pervious sand. Drainage blankets were provided near the toe, between the dam's impervious fill material and the dam's foundation soils, to control seepage. The seepage from the dam becomes excessive at certain locations of the dam; therefore, it is necessary to identify the seepage flow paths early in the investigation.

Fujimoto (2009) performed a study on the application of resistivity imaging (RI) to determining seepage flow paths at the Lewisville Dam. The seepage area of the dam is presented in Figure 6.37.

The Lewisville Dam was constructed with impermeable earthen materials on top of bedrock, impervious, semi-impervious, and pervious soils. The soil profile before the construction of the dam is shown in Figure 6.38. Figure 6.39 shows the soil profile integrated by several borings along the toe from Station 78+50 to Station 82+00. Based on the soil profiles, a sedimentary rock layer or shale underlies a sand layer and a clay layer. The shale is dark gray to yellow brown and is slightly weathered. The sand layer above the shale layer is medium dense to loose and saturated. The color of the sand is described as yellow brown to light gray. There is a yellow brown to gray, moist, and sandy clay layer above the sand layer.

The resistivity imaging was conducted at seepage area #1 (Figure 6.37). The area covers 1100 ft. of the embankment; the station numbers for the investigation are shown in

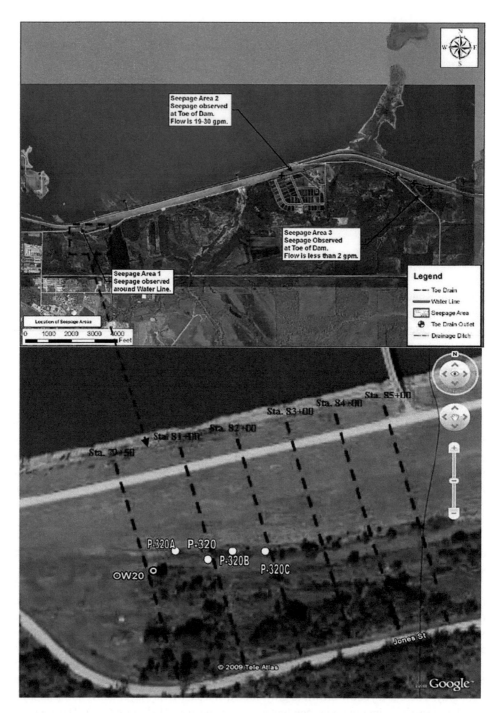

Figure 6.37 Location of the seepage area of the Lewisville Dam (Fujimoto, 2009)

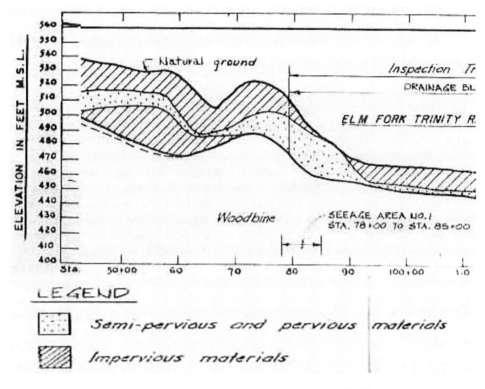

Figure 6.38 Soil profile before the dam construction (Fujimoto, 2009)

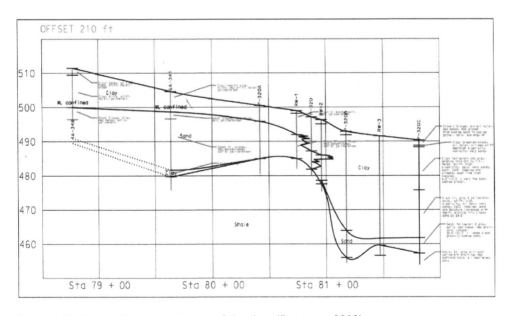

Figure 6.39 Soil profile along the toe of the dam (Fujimoto, 2009)

Figure 6.37. Piezometers and observation wells were installed near the toe of the dam. Fujimoto (2009) selected the area from Station 79+50 to Station 85+00 to perform the resistivity imaging. The embankment has an approximate 3.5 horizontal to 1 vertical slope over the seepage area #1. The elevation of the top of the embankment is 560 ft., the elevation of the toe of the downstream side is approximately 500 ft., and the average lake water level is 520 ft.

Fujimoto (2009) conducted seven parallel RI tests. The layout of the 2D RI tests are presented in Figure 6.40. Dipole-dipole array was used with an electrode spacing of 3 m (10 ft.) for each testing. In addition, the roll-along technique was utilized in some of the testing to increase the array length. Field setup for the RI tests is presented in Figure 6.41.

Based on Fujimoto's (2009) work, the RI investigation results of seven testing lines are presented in Figure 6.42. Resistivity values less than 8 ohm-m are interpreted as possible seepage flow paths, and the light blue color denotes 8 ohm-m resistivity. Figure 6.43 shows only low resistivity areas from Line 1 through Line 6, and it helps to see the possible seepage flow paths. The 2D RI profile along Line 1 indicated low resistivity area between Station 80+10 and 83+76 at the elevations from 490 ft. to 470 ft. It is also evident from the RI results that the thickness of the permeable layer is approximately 25 ft. In Line 1, from Station 83+00 to 85+00, there is high resistivity area which may be interpreted as an impermeable layer. Line 2 to Line 7 started from Station 79+50. The 2D RI test results consistently indicated a low resistivity area between Stations 80+50 and 82+84. Moreover, the thickness of the low resistivity area (probable seepage area) was observed within 27–29 ft.

Fujimoto (2009) also compared existing field piezometer results with 2D RI test results. There are three piezometers along the toe of the dam, at 80+93 (P-320), 81+43 (P-320B), and 81+93 (P-320C). The data of the piezometers is shown in Figure 6.44. These three piezometers are located on Line 1, and the data was used to crosscheck the validity of the RI profiles

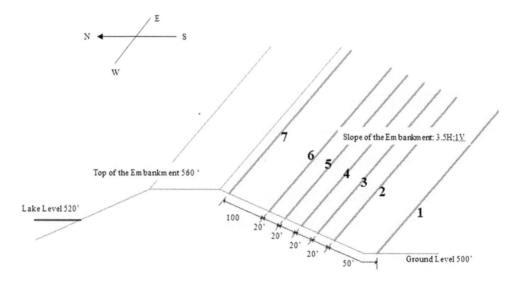

Figure 6.40 Testing lines on the embankment (Fujimoto, 2009)

Figure 6.41 Field setup of RI tests (Fujimoto, 2009)

on Line 1, which is located along the toe. Table 6.1 summarizes the data from the piezometers and readings from the RI profile on Line 1.

Fujimoto (2009) utilized the low resistivity values in Line 1 to Line 7 and developed a possible resistivity profile through the embankment (Figure 6.45). The possible seepage area was also compared to the soil profile, as shown in Figure 6.38. It was observed that the low resistivity area, or probable seepage flow paths, are in the subsurface beneath the area of the investigation. Seepage beneath the dam seems to flow through the native pervious and semi-impervious earthen materials, between the dam's impervious fill material and the Woodbine bedrock formation (shale). It is the native formation of the subsurface, where seepage is able to flow out of the reservoir because the most probable conductors in the subsurface earthen materials are the groundwater flow paths, which are the most saturated zones. Figure 6.46 shows probable seepage paths on the soil profile.

Proper identification of the seepage area is crucial to maintaining the safety of the structure. Soil borings and installation of the piezometer aid in identifying any potential for seepage; however, it is difficult to locate the overall extent of the seepage area by using soil boring. Moreover, many dam owners are reluctant to drill through the earth dam for investigative purposes. Since the 2D resistivity imaging is a non-invasive method, it can be utilized to locate the extent and depth of the possible seepage areas in earth dam.

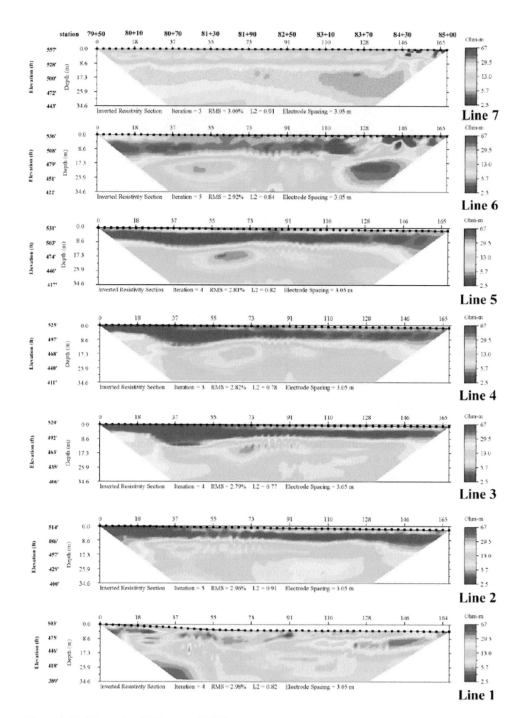

Figure 6.42 RI results (Fujimoto, 2009)

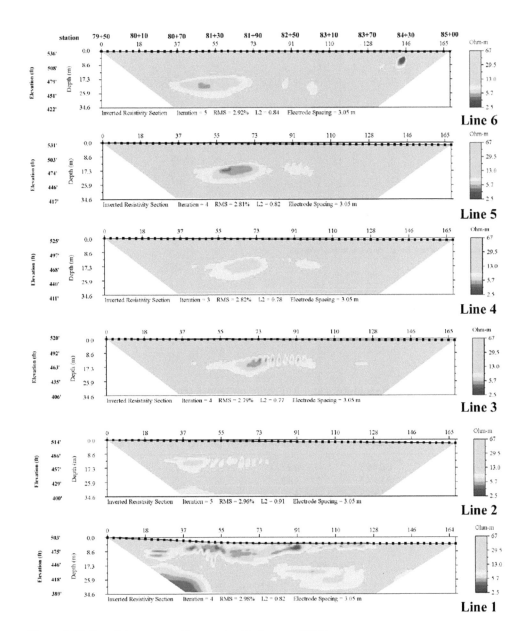

Figure 6.43 Possible seepage flow paths (Fujimoto, 2009)

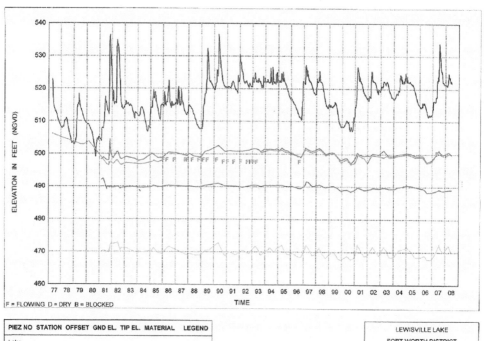

PIEZ NO	STATION	OFFSET	GND EL.	TIP EL.	MATERIAL	LEGEND
Lake						
P-320A	80+43	220 D/S	500.20	486.00	OVB(Sd)	
P-320	80+93	221 D/S	496.70	481.20	OVB(Sd)	
P-320B	81+43	220 D/S	492.80	460.40	OVB(Sd)	
P-320C	81+93	220 D/S	490.40	457.30	OVB(Sd)	

LEWISVILLE LAKE
FORT WORTH DISTRICT
PIEZOMETERS
P-320A, P-320, P-320B, P-320C
PIEZOMETRIC ELEVATION vs. TIME

Figure 6.44 Data from the piezometers at the Lewisville Dam (Fujimoto, 2009)

Table 6.1 Data from piezometers and RI profile on Line 1 (Fujimoto, 2009)

	Piezometer		RI Profile on Line 1
	Station	Piezometric Height (ft)	Possible Phreatic Line
P320	80+93	497	489
P320B	81+43	488	486
P320C	81+93	470	475

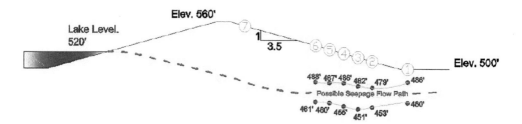

Figure 6.45 Possible seepage flow through embankment (Fujimoto, 2009)

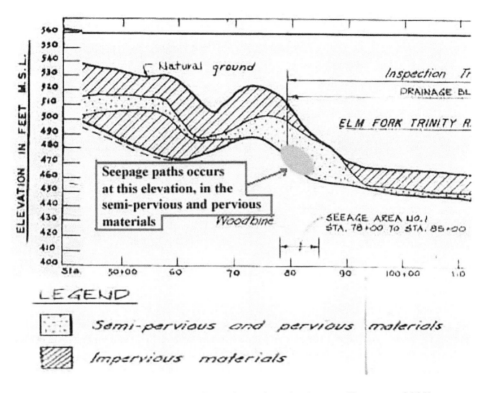

Figure 6.46 Cross section of seepage flow through embankment (Fujimoto, 2009)

6.5 Application of RI for the Evaluation of Unknown Foundations

Bridges are classified as having unknown foundations when the type, dimensions, reinforcement, and/or elevations are unknown. About 90,000 bridges with unknown foundations have been identified by the National Evaluation Program (Richardson and Davis, 2001). Unknown foundations pose significant problems and are an important safety concern for transportation agencies. Th evaluation of existing foundations differs from the usual nondestructive testing (NDT) method in which structures cover the top of deep foundations (Gassman and Finno, 1999). Several methods are available for the evaluation of unknown foundation depths.

Hossain *et al.*, 2011 used RI to conduct a study to determine the depth of a driven pile foundation of a bridge. The bridge was located at FM 2738 over Mountain Creek in Fort Worth, Texas. A 2D RI was conducted along the bridge bents at dry Mountain Creek. The bridge was supported by four steel H-piles (P1, P2, P3, and P4). The depth of the foundation was known to the agency; however, an RI test was conducted to check the feasibility of this technique. A total of 56 electrodes with an electrode spacing of 0.6 m (2 ft.) were utilized in RI. The total array length of the RI was 33.5 m (110 ft.). The RI layout and field setup are presented in Figure 6.47 and Figure 6.48, respectively.

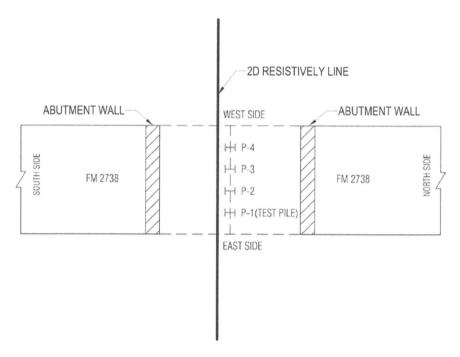

Figure 6.47 Resistivity layout for bridge foundation at Mountain Creek on FM 2738 (Hossain et al., 2011), with permission from ASCE

Figure 6.48 Resistivity setup for Mountain Creek on FM 2738 (Hossain et al., 2011), with permission from ASCE

The RI test result is presented in Figure 6.49. The resistivity profiles below the pile P1 to P4 are presented in Figure 6.50.

Based on the resistivity results as presented in Figure 6.49, a zone of high resistivity materials was observed, and the unknown foundation depths were determined from it. During the installation of driven pile into this bridge site, the H-pile was driven into the

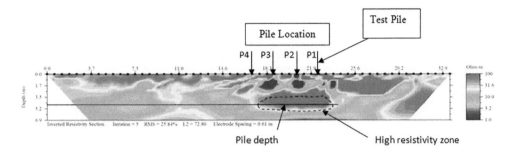

Figure 6.49 Resistivity imaging results on bridge foundation at Mountain Creek over FM 2738 (Hossain *et al.*, 2011), with permission from ASCE

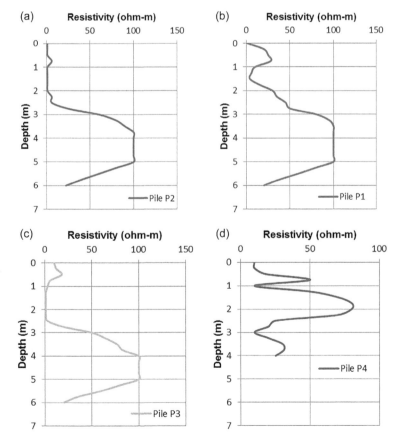

Figure 6.50 Variations of resistivity under bridge foundation at Mountain Creek at FM2738: (a) Pile P1, (b) Pile P2, (c) Pile P3, and (d) Pile P4

foundation soil. As a result, a disturbed zone or compacted soil zone was expected to be present immediately below the pile tip. From the variations of resistivity, as presented in Figure 6.50 (a) to Figure 6.50 (c), an increase in resistivity was observed from 3.5 m to 5 m (11.5 to 16.4 ft.) below the pile. The pile foundation was estimated as 4.35 m (14.25 ft.), which was the average of the high resistivity areas. From Figure 6.49 and Figure 6.50 (d), it was observed that the high resistivity zone had a foundation depth of 2 m (6.5 ft.) below the ground level. Thus, according to RI, the pile length for pile P4 was shorter than that of the other three piles.

Pile P1 was also tested, using the parallel Seismic and Sonic echo methods to verify the RI results. The parallel seismic method provided results similar to those obtained from RI for pile P1; however, the sonic echo method overestimated the pile depth. According to the installation record, the actual depth of P1, P2, and P3 was 4.3 m (14 ft.), and the depth of P4 was 2.1 m (7 ft.).

The use of RI in evaluating unknown bridge foundations is comparatively new and is more limited than other NDT techniques. However, several researchers are studying this technique to investigate the type and depth of the unknown bridge foundations.

6.6 Application of RI to investigate moisture variations in bioreactor landfills

The bioreactor landfill is a comparatively new concept. It was designed and is operated for rapid municipal solid waste (MSW) decomposition, enhanced gas production, and waste stabilization. In favorable environmental conditions, biological stabilization of the waste in a bioreactor landfill is faster than in a conventional or dry landfill. The rate of solid waste decomposition significantly depends on the moisture content of the waste. Moisture distribution and the extent of leachate recirculation are the primary concerns when attempting to optimize the performance of a bioreactor landfill. A leachate recirculation system installed in a bioreactor landfill is designed to distribute the moisture uniformly throughout the landfill. However, due to high heterogeneity and anisotropy of MSW and the different compaction levels in landfill, the moisture distribution may not be uniform throughout the entire landfill. RI techniques have been utilized by many researchers to investigate the moisture variations in bioreactor landfills. Two of the case studies are presented in the following sections.

6.6.1 Investigation of moisture variation using RI

Manzur *et al.*, 2016 extensively investigated the moisture variations of the recirculation system of a bioreactor landfill in Denton, Texas. A horizontal recirculation pipe was selected within Cell 2A of the Denton Landfill and monitored for 2.5 years. A total of 36 pipes were installed in Cell 2A, of which 24 pipes had special waste and 12 pipes had tire chips as backfill material. Pipe H2 was selected for studying the moisture variations and movement during leachate recirculation. The horizontal pipe H2 had a 150 mm (6 inch) diameter HDPE pipe, with 33.1 m (100 ft.) solid sections at both the ends and a perforated section in the middle. It was backfilled with special waste. The layout and cross section of the horizontal recirculation pipe in Cell 2A is presented in Figure 6.51.

Manzur *et al.* (2016) performed RI tests at different times, along and across the recirculation pipe H2 after leachate recirculation, to map the effects of leachate recirculation or the addition of moisture into the solid waste mass. The RI tests were conducted using

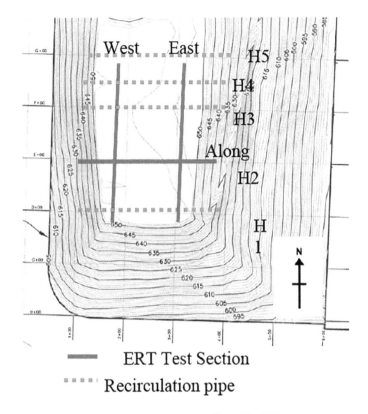

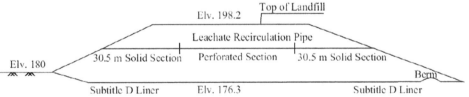

Figure 6.51 Landfill cross section of leachate recirculation pipe H2 (Manzur *et al.*, 2016)[7]

56 electrodes, with a programmable eight-channel, Super-Sting R8/IP instrument. The electrodes were placed at 1.83 m (6 ft.) center-to-center intervals, resulting in a test line of 100 m (330 ft.). Dipole-dipole array was utilized during the RI, as it provides better resolution in both horizontal and vertical directions.

Manzur *et al.* (2016) conducted the RI test before the addition of any liquid, and this was considered the "baseline" or "reference" test. The baseline RI was performed on 8 May 2009, along recirculation pipe H2. Later, RI tests, referred to as recirculation tests, were conducted

7 Reprinted from Waste Management,, 55, Manzur, S. R., Hossain, M. S., Kemler, V., & Khan, M. S. , Monitoring extent of moisture variation due to leachate recirculation in an ELR/bioreactor landfill using Resistivity Imaging, 8-48, Copyright (2016), with permission from Elsevier.

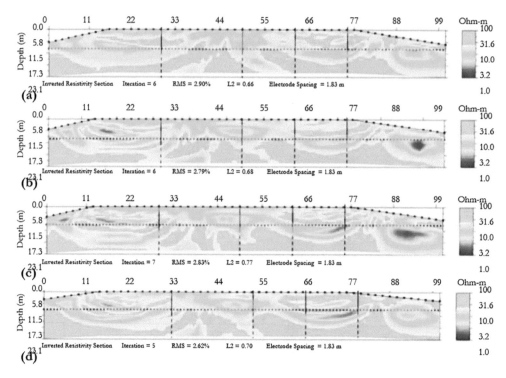

Figure 6.52 Monitoring leachate accumulation along pipe H2: (a) Baseline RI on 05/08/2009, (b) Recirculation RI test after 1 day of addition of 15180 liters (4,010 gallons) on 07/10/2009, (c) Recirculation RI test after 1 day of addition of 31,300 liters (8,268 gallons) on 06/24/2010, and (d) Recirculation RI test after 1 day of addition of 41670 liters (11,007 gallons) on 02/16/2011 (Manzur et al., 2016)[8]

on the same test section after adding liquid during the regular operational period. A series of RI tests were performed on pipe H2 after leachate recirculation. The test results are presented in Figure 6.52.

The recirculation test along pipe H2 was performed on 10 July 2009, when 15,180 liters (4010 gallons) of leachate were added through pipe H2. The RI test results, presented in Figure 6.52, illustrate a reduction in resistivity at several locations below the recirculation pipe. The qualitative diagrams show a reduced resistivity contour below pipe H2, representing the flow of leachate after leachate recirculation through the pipe. The 2D electrical resistivity profile also indicated the varying resistivity distribution that confronted non-uniform leachate flow throughout the pipe. Leachate was recirculated by a City of Denton official again on 23 June 2010, when 31,300 liters (8268 gallons) of leachate were added into pipe H2. Mazur et al. (2016) conducted the corresponding recirculation test on 24 June 2010. Another RI test was conducted after recirculating 41,670 liters (11,007 gallons) of water on

8 Reprinted from Waste Management,, 55, Manzur, S. R., Hossain, M. S., Kemler, V., & Khan, M. S. , Monitoring extent of moisture variation due to leachate recirculation in an ELR/bioreactor landfill using Resistivity Imaging, 8-48, Copyright (2016), with permission from Elsevier.

16 February 2011. The study indicated, from continual monitoring from 2009 to 2011, that the cumulative addition of water continuously reduced the resistivity of the underlying waste.

In addition to the RI test along the recirculation pipe, Manzur *et al*. (2016) investigated the extent of the moisture variations after leachate recirculation. A test section located towards the west side slope at Cell 2A was selected for investigating the moisture distribution across pipe H2. The layout of the west test section is presented in Figure 6.51. The baseline RI tests were performed at the west section in 22 May 2009. The west test section was located along grid line 2+00. The test section was selected from C+90 up to G+20 that covered a total section of 100 m (330 ft.). The leachate was injected near the western slope. Leachate recirculation tests were carried out for pipe H2 on the day following the addition of liquid. The RI test results are presented in Figure 6.53.

The baseline resistivity image for the tests performed across pipe H2 is presented in Figure 6.53a. The resistivity test was conducted at the same location as the baseline test; therefore, it was expected that the resistivity value would be the same if no moisture was injected. The addition of moisture increases the conductivity and decreases the electrical resistance. Thus, the zone of low resistivity contour around the recirculation pipe was considered to be the zone of influence across the pipe which directly corresponded to the quantity of the injected leachate. The variation in the zone of influence with time can be observed in Figure 6.53 (a) through Figure 6.53 (g). With the addition of water or leachate, the gray area, representative of high resistance, below the leachate recirculation pipe decreased slowly, as presented in Figure 6.53 (a) to Figure 6.53 (c). This indicates that moisture travelled downwards with each additional leachate recirculation or moisture addition.

Manzur *et al*. (2016) facilitated the 2D ERI results to investigate and quantify the extent of recirculation, on both the right and left sides across pipe H2. To do that, a 30 m (100 ft.) section was selected from the 2D resistivity imaging results, and the resistivity value was extracted every 3.66 m (12 ft.). Therefore, the locations of the changes in resistivity were monitored at 3.66 m (12 ft.), 7.32 m (24 ft.), 10.98 m (36 ft.), and 14.63 m (48 ft.) from the pipe on both the left and right sides of pipe H2. The schematic vertical planes across pipe H2 and the changes in resistivity at different times along the vertical planes are presented in Figure 6.54.

The vertical electrical resistivity profiles in Figure 6.54 indicate that the baseline resistivity ranged from 20 Ω.m to 50 Ω.m for all four selected locations on the left side of Pipe H2. However, the resistivity was observed to be slightly lower, ranging from 10 Ω.m to 35 Ω.m, at all four selected locations toward the right side of pipe H2. The non-homogeneity and anisotropy of the solid waste materials may have influenced the differential resistivity ranges on both sides of the pipe. Based on the study, the resistivity profile followed an inward movement that depicted the reduction in resistivity with leachate recirculation. During the initial stages of recirculation, the reduction of the resistivity was fairly tepid; however, with the ongoing recirculation, the resistivity significantly decreased at all four locations on the left side of the pipe. The resistivity values dropped from 50 Ω.m to20 Ω.m at 3.66 m (12 ft.) to 14.63 m (48 ft.) on the left side of the pipe H2 during the monitoring period. Likewise, the change of resistivity on the right side of the pipe, as presented in Figure 6.54, followed a similar trend. The reduction in resistivity was slow during the initial stages; however, the resistivity decreased from 40 Ω.m to 25 Ω.m at (b) 3.66 m, (c) 7.32 m, (d) 10.98 m over the entire monitoring time period. The resistivity reduced as much as 45% to 60% after leachate recirculation. In contrast, the resistivity increased 14.64 m (48 ft.) to the right of Pipe H2, indicating that moisture did not travel as much as 14.64 m (48 ft.) on the right side of the

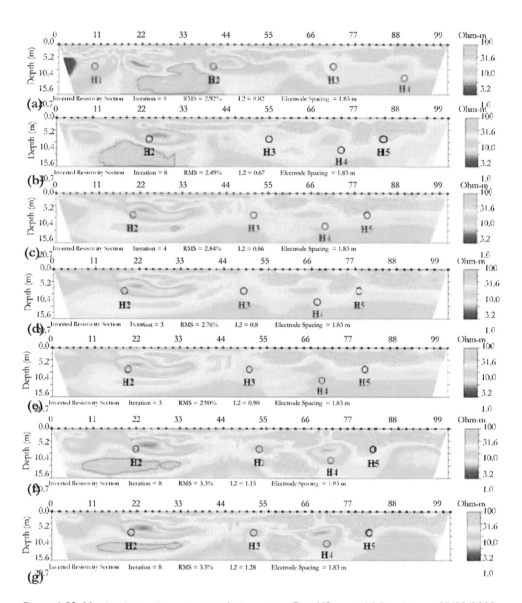

Figure 6.53 Monitoring moisture accumulation across Pipe H2 west: (a) Baseline on 05/22/2009, (b) 1 day after recirculation of 18,920 liters (5006 gallons) on 7/29/2009, (c) 1 day after recirculation of 19,470 liters (5151 gallons) on 8/26/2009, (d) 1 day after recirculation of 18,940 liters (5010 gallons) on 9/26/2009, (e) 1 week after recirculation of 18,940 liters (5010 gallons) on 10/2/2009, (f) 1 week after recirculation of 37,870 liters (10019 gallons) on 10/23/2009, and (g) 1 week after recirculation on 22,715 liters (6009 gallons) on 11/4/2009 (Manzur et al., 2016)[9]

9 Reprinted from Waste Management,, 55, Manzur, S. R., Hossain, M. S., Kemler, V., & Khan, M. S. , Monitoring extent of moisture variation due to leachate recirculation in an ELR/bioreactor landfill using Resistivity Imaging, 8-48, Copyright (2016), with permission from Elsevier.

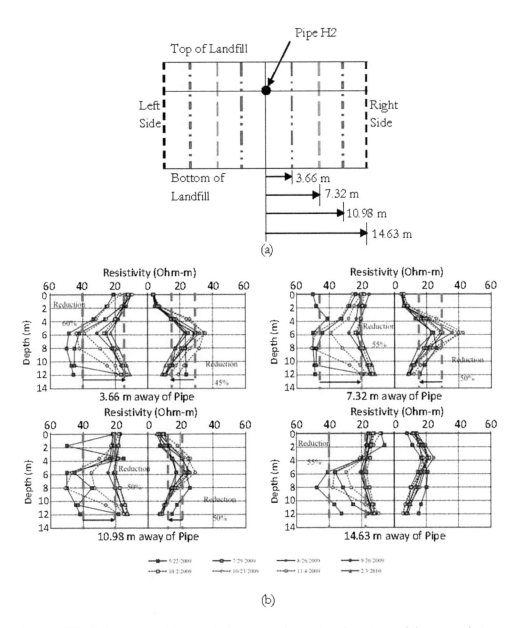

Figure 6.54 (a) Schematic of the vertical plane to determine the extent of the recirculation (b) Change of resistivity with recirculation of pipe H2 at west section (Manzur *et al.*, 2016)[10]

10 Reprinted from Waste Management,, 55, Manzur, S. R., Hossain, M. S., Kemler, V., & Khan, M. S. , Monitoring extent of moisture variation due to leachate recirculation in an ELR/bioreactor landfill using Resistivity Imaging, 8-48, Copyright (2016), with permission from Elsevier.

pipe after recirculation. Manzur *et al.* (2016) concluded that the extent of recirculation was determined to be approximately 14.5 m (47.5 ft.) on the left side and 11.0 m (36 ft.) on the right side of the recirculation pipe H2.

6.6.2 Application of RI in performance evaluation of a bioreactor landfill operation

Alam *et al.*, 2017 conducted a study to monitor the performance and evaluate the efficiency of a bioreactor landfill operation based on different operational parameters. During the study, Alam *et al.*, 2017 utilized the electrical resistivity imaging technique and ground subsidence data to evaluate the performance. The study was primarily conducted in Cell 2 of the City of Denton Landfill, in Denton, Texas.

A total 36 horizontal recirculation pipes are located at Cell 2 at the City of Denton Landfill to recirculate water or leachate into the landfill. For the purpose of the study, Cell 2 was divided into three zones: zone 1, zone 2, and zone 3. Figure 6.55 depicts the locations of the horizontal pipes, where H2 and H16 pipes were selected based on the maximum recirculation through each pipe, to monitor moisture distribution, using the electrical resistivity imaging (ERI) method. No recirculation pipes were selected from zone 3. Lines A-A, B-B and C-C in Figure 6.55 represent the field ERI test lines.

Moisture distribution due to leachate recirculation through horizontal pipes was monitored at the City of Denton Landfill to assess its performance from May 2009 to April 2012.

Alam *et al.*, 2017 performed the RI test across each pipe before leachate recirculation in order to obtain baseline data, as shown in Figure 6.55. ERI tests were also performed one day, one week (seven days), and two weeks (14 days) after leachate recirculation to observe moisture distribution with time, as presented in Figure 6.56. Observations near pipe H2 in the baseline study showed a significantly high resistivity gray zone. The ERI results, after one day of leachate recirculation, indicated the presence of a low resistivity blue zone and a decrease of the high resistivity area. The low resistivity zone expanded more after seven days of leachate recirculation, with a decrease in the high resistivity area. The ERI results after 14 days of leachate recirculation showed moderate resistivity values, with mostly green and yellow areas.

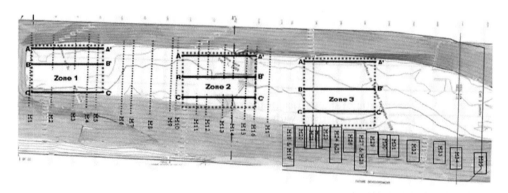

Figure 6.55 Horizontal pipes and zones at Cell 2 (Alam *et al.*, 2017), with permission from ASCE

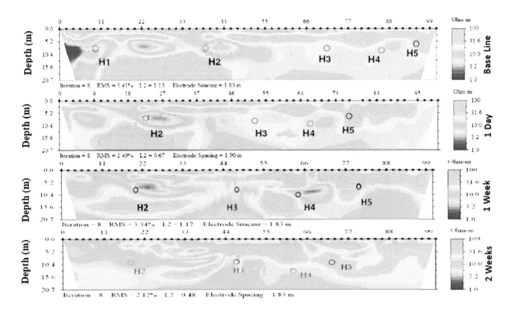

Figure 6.56 Resistivity imaging results of pipe H2 for first year (May 2009 to April 2010) (Alam *et al.*, 2017), with permission from ASCE

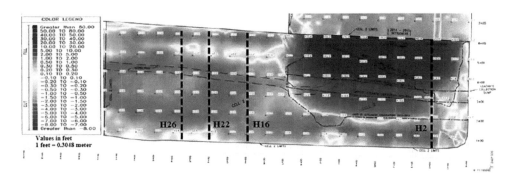

Figure 6.57 Landfill subsidence from 2010 to 2015 at Cell 2 (Alam *et al.*, 2017)., with permission from ASCE

The settlement of the City of Denton Landfill has been conducted by an annual subsidence survey since January 2010. Figure 6.57 represents the subsidence of the landfill at Cell 2 for the years of 2010 to 2015.

Alam *et al.*, 2017 indicated that landfill settlement increased with time, from 0.03 m to 1.01 m within five years of bioreactor operation, as presented in Figure 6.58 (a). Similar trends were observed for the subsidence data across recirculation pipes, where settlement ranges from 0.03 m to 1.43 m, were observed, as presented in Figure 6.58 (b). This similarity represents horizontal movement of added water/leachate within the waste mass. Minimal non-uniform settlement of the MSW along and across recirculation pipes indicates the heterogeneity of waste material. Both results, along and across, indicate that

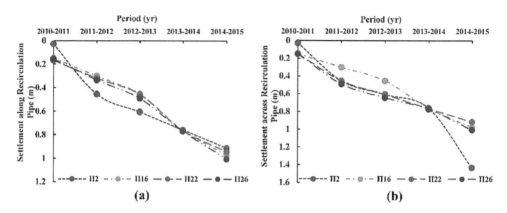

Figure 6.58 Changes in elevations: (a) along recirculation pipes of Cell 2, and (b) across recirculation pipes of Cell 2 (Alam *et al.*, 2017), with permission from ASCE

the settlement increased each year and demonstrate that the bioreactor landfill operation was working efficiently. Additional space was gained for each period from 2010–2011 to 2014–2015 and was collected from the City of Denton Landfill. The additional space gain in Cell 2 increased significantly due to faster waste degradation. Alam *et al.*, 2017 also indicated that the cumulative additional space gained due to the bioreactor operation over the course of five years amassed a total of 77,000 m³. Air space recovery is one of the major advantages of bioreactor landfills because of the potential utilization of space for future landfill operations.

The ERI test results helped to understand the overall moisture distribution within the landfill. The decomposition of the MSW in the bioreactor cell of City of Denton Landfill increased significantly, over the past five years, due to the distributed leachate recirculation, and resulted in subsidence and an air space gain. Thus, it can be concluded that the ERI is an integrated tool that is effective for monitoring the moisture variations and evaluating the performance of bioreactor cells.

6.7 Application of RI to investigate moisture variations of expansive subgrades

Expansive soils are highly plastic in nature and undergo volumetric deformation due to seasonal moisture variations. Consequently, lightweight structures such as highway pavements and foundations built on top of expansive soil show distress a few years after construction. Identification of moisture variations under the pavement on expansive subgrades is an important task to determine the cause(s) of pavement distress.

A geophysical investigation was conducted, using RI, to determine the moisture movement in different seasons at the slope of the pavement in SH 342 in Lancaster, Texas (Figure 6.59). A subsurface investigation program indicated that it has highly plastic clay soil (CH) up to a depth of 15 ft., which experiences shrink-swell behavior with seasonal moisture variations. The liquid limit of the soil varied between 55 and 70, and the plasticity index of the soil samples varied between 35 and 45. The layout of the 2D RI line is presented

Figure 6.59 Resistivity line in SH 342

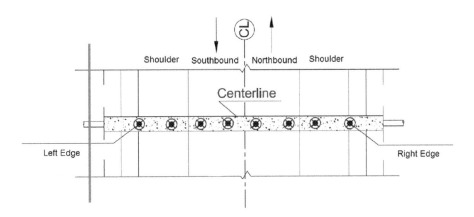

Figure 6.60 Location of resistivity imaging

in Figure 6.60. The RI test was conducted along the edge of the pavement every month to observe the seasonal moisture variations beneath the pavement.

The resistivity plots for the months from October 2015 to July 2016 are shown in Figure 6.61. It can be observed from the figure that the plots have lower resistivity from October to April, when the wet season occurs. The plots for May 2016 to July 2016 reflect higher resistivity, indicating the dry season. In the plot of May 2016, a low resistivity zone can be observed at the center, at around 3 ft., indicating the presence of moisture. Resistivity

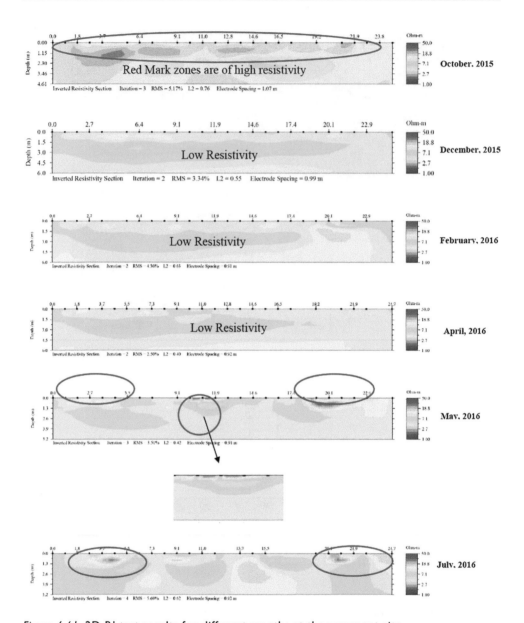

Figure 6.61 2D RI test results for different months at the pavement site

conducted in May 2016 followed a rainfall event, which revealed an edge crack by which moisture entered the pavement.

Similar trends were observed throughout the 2016–2017 monitoring period. November to April was found to be wetter, while May to October was dryer, confirming the seasonal variations of moisture in the pavement, as presented in Figure 6.62.

Further analysis was conducted from the 2D RI test results to determine the active zone and eventually understand the seasonal variations. The active zone is the depth of the seasonal moisture variation zone, which also undergoes volumetric deformation of the soil. The

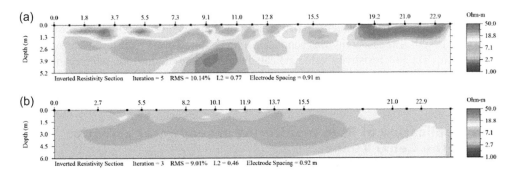

Figure 6.62 (a) Typical dry period (2016–17), and (b) typical wet period (2016–17)

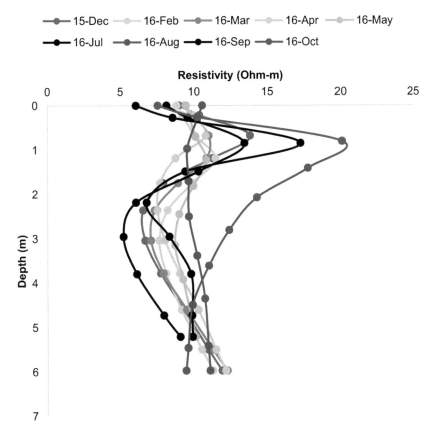

Figure 6.63 Change of resistivity with depth

resistivity variations along the midpoint of the study are presented in Figure 6.63. It can be seen from Figure 6.63 that changes of resistivity values followed a narrow band (2–2.5 m) after a depth of 3.1 m (11 ft.). Hence, it can be concluded that changes in moisture are not that prominent after this depth, and it can be considered the active zone, after which changes in moisture are not significant.

As can be seen from the figure above, the resistivity values of the active zone are within the 7–13 Ohm range in the dry season; in the wet season, they reach up to 23 Ohm. Another interesting fact, depicted in Figure 6.62, is that resistivity as low as 5 Ohm was observed at 1 m (3 ft.) depth. The resistivity was conducted after a rainfall event on that particular day (26 May 2016). The low resistivity value indicated the presence of moisture, which can be also seen in Figure 6.63. Hence, resistivity imaging can also capture possible sources of moisture intrusion from the edge.

6.8 Application of RI in monitoring groundwater activities at the Cenxi Tunnel

Lui *et al.*, 2017 conducted a study to investigate water inrush at the Cenxi Tunnel in the Guangxi Province of China. A severe water inrush occurred in the tunnel, resulting in many environmental geological disasters, such as water loss in the nearby lakes, wells, dried-up springs, and subsidence on the ground surface. Lui *et al.*, 2017 investigated the key groundwater runoff with multiple surface RI survey lines. A 2D cross-hole ERT survey was then conducted to monitor the variations of groundwater during the tunnel water inrush and the grouting process that followed. The location and photo of the water inrush are presented in Figure 6.64.

The survey field was located in the tectonic erosional mountains of southeastern Guangxi, China, where the terrain is undulate and has V-shaped gullies and steep hillsides due to the long-term tectonic effect and surface water erosion. The tectonics are where the rock mass of inner bedrock developed with joints, resulting in cracks and partially crushed rock. Water-conducting pathways are most likely to develop under these hydraulics impacts. Within a short period after the water inrush, surface subsidence occurred at several locations on the west side, above the tunnel face, in a range of about 800 m. The preliminary geological investigation indicated at least one underground runoff that ran from west to east.

Lui *et al.*, 2017 arranged four surface ERT survey lines, the layout of which is presented in Figure 6.64. The stretch of the survey lines was vertical to the connecting line between the tunnel face and surface subsidence. The survey lines were numbered from 1 to 4, from east to west. Survey lines 1, 2, and 4 were 324 m long, with electrode spacing of 6 m. Survey Line 3 was 132 m, with electrode spacing of 3 m. The Wenner array was used to collect field data. The 2D RI results of the four lines, arranged according to the real spatial orientation and the corresponding location of the survey lines, are presented in Figure 6.65. The 2D RI test results revealed a very heterogeneous terrain. The background material had a resistivity value greater than 600 Ωm, with > 4000 Ωm indicating a high resistivity area. It should be noted that several low resistivity zones can be found in the inversion result of each survey line, indicating that the rock mass in these areas is crushed and filled with water. A continuous low resistivity area was located at the surface of survey Line 4, with this conductive area corresponding to the presence of a nearby irrigation canal. Lui *et al.*, 2017 identified a significant continuous conductive area as the preferred runoff pathway of groundwater, which is marked with black ellipses in Figure 6.72. From the images, it was determined that the buried depths are 12–30 m (Line 4), 15–30 m (Line 3), 35–80 m (Line 2), and 80–108 m (Line 1). The runoff pathway orients downward from west to east, as shown in Figure. 6.72.

Lui *et al.*, 2017 conducted a further field experiment to understand the impact of the tunnel water inrush on the groundwater environment by monitoring the groundwater migration in the runoff pathway found on survey Line 1 (Figure 6.65). Two 70 m boreholes were drilled

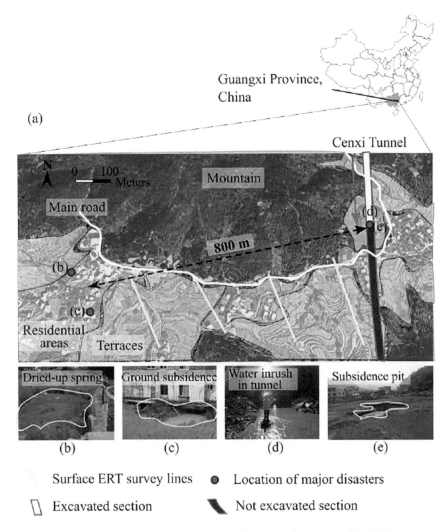

Figure 6.64 Geographical and geological maps of the studied area: (a) field overview and arrangement of survey lines, (b) photo of the dried-up spring, (c) photo of ground subsidence around a private house, (d) photo of water inrush in tunnel, and (e) photo of ground subsidence pit right above the face of Cenxi Tunnel (Liu *et al.*, 2017)[11]

across the water-conducting area deduced on survey Line 1, with a distance of 26 m between them (Figure 6.66).

As presented in Figure 6.66, the depth of the water-bearing zone in the left borehole coincided well with that in the inversion image. Using ERT, Lui *et al.*, 2017 monitored a 2D

11 Reprinted from Tunnelling and Underground Space Technology, Vol 66, Bin Liu, Zhengyu Liu, Shucai Li, Kerui Fan, Lichao Nie, Xinxin Zhang, An improved Time-Lapse resistivity tomography to monitor and estimate the impact on the groundwater system induced by tunnel excavation, 107-120, Copyright (2017), with permission from Elsevier.

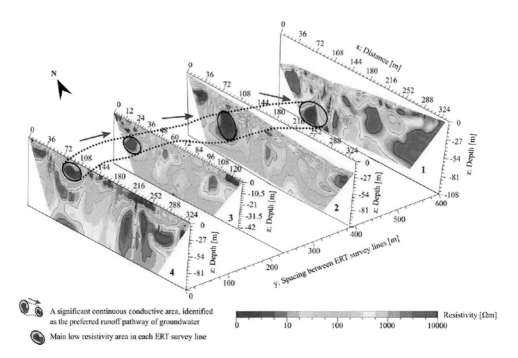

Figure 6.65 Inversion results of surface ERT measurements for survey lines 1 to 4 and the underground runoff pathway from west to east (Liu et al., 2017)[12]

cross-hole with an electrode spacing of 2 m and a length of 70 m in boreholes #1 and #2. The dipole-dipole array was used in the continuous monitoring process of the water-conducting area. The monitoring process lasted two weeks (from 14 to 30 Sept.) and was divided into three stages. In the first stage, a large quantity of water (600–800 m3/h) continuously gushed out of the tunnel. At the second stage, curtain grouting was made to stop the water (water inrush was less than 100 m³/h) after a subsidence pit occurred above the tunnel face. In the third stage, the water inrush became stable (at 100–200 m³/h) after three consecutive days of rain. Lui et al., 2017 selected five moments (14 Sept., 18 Sept., 22 Sept., 26 Sept., and 30 Sept.) to collect the data. The inversion result of the initial dataset which was acquired on 14 Sept. (T1) is shown in Figure 6.67a. As the initial inversion model, Liu et al., 2017 used a homogeneous apparent resistivity model with a smooth factor of $\lambda = 0.22$, eight iterations and a convergence error RMS that satisfied the $< 3\%$ requirement. The test results indicated that the elevation of the groundwater in the water-conducting area was 78 m below the surface.

The datasets of T2 – T5 were normalized by the same ratio as the model at T1, as an initial reference model for inversion. Afterwards, the normalized datasets at T2 – T5 were substituted into the time-lapse inversion. It should be noted that the primary focus of the 2D cross-hole ERT was to investigate the vertical migration characteristics of the groundwater

12 Reprinted from Tunnelling and Underground Space Technology, Vol 66, Bin Liu, Zhengyu Liu, Shucai Li, Kerui Fan, Lichao Nie, Xinxin Zhang, An improved Time-Lapse resistivity tomography to monitor and estimate the impact on the groundwater system induced by tunnel excavation, 107-120, Copyright (2017), with permission from Elsevier.

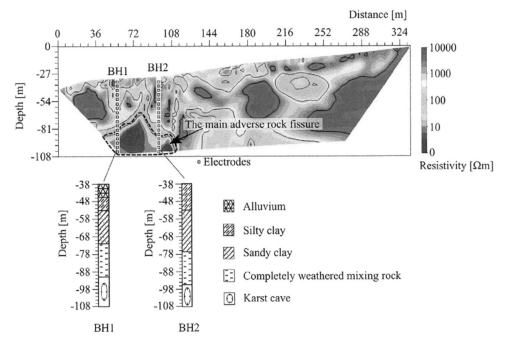

Figure 6.66 Inversion result of surface ERT Line 1, the locations of the boreholes (BH1 and BH2), and their core drilling lithology (Liu *et al.*, 2017)[13]

in the runoff channel. Parameters of the temporal weighted matrix were taken as $\mu = 0.5$, $\eta = 1.0$, $\alpha = 0.2$, and $\gamma = 0.4$, with eight inversion iterations for each time step. Images showing the percentages of change in resistivity in T2 – T5 are shown in Figure. 6.74 (b – e).

Liu *et al.*, 2017 indicated that the temporal variations of resistivity were not significant in the upper part of the model (−38 m to −68 m in Figure.6.74). However, obvious changes appeared in the lower part of the 2D cross-hole ERT. The change in the percentage was from −50% to 50% from the first survey to the last one. Based on the results of the independent inversion of the initial dataset, Lui *et al.*, 2017 interpreted the lower part of the model with a low resistivity area (< −78 m) as a water-bearing structure. The positive change of resistivity in TL-ERT results (H1 and H2) translate to a loss of water, and the negative change corresponds to the water recharge processing.

Based on Figure 6.67(b) and Figure 6.67(c), an increasing resistivity area is under −78 m, translating to a trend of downward evolution over time (from H1 to H2), and indicated loss of groundwater with the high flow rate of the water inrush in the tunnel. Lui *et al.*, 2017 indicated that the elevation of the groundwater decreased 14 m from the depth level of −78 m to −92 m from 14 Sept. to 18 Sept., and it continued to decrease from −92 m to −102 m from 18 Sept.

13 Reprinted from Tunnelling and Underground Space Technology, Vol 66, Bin Liu, Zhengyu Liu, Shucai Li, Kerui Fan, Lichao Nie, Xinxin Zhang, An improved Time-Lapse resistivity tomography to monitor and estimate the impact on the groundwater system induced by tunnel excavation, 107-120, Copyright (2017), with permission from Elsevier.

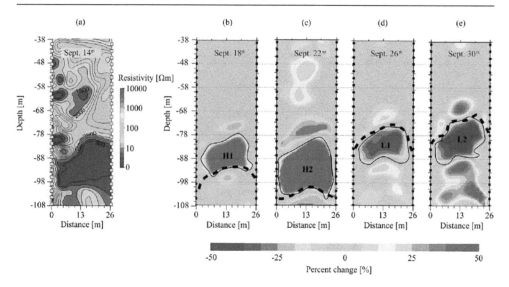

The initial state of groundwater in fractured rock mass.

The variation of groundwater at different monitoring time step.

H1, H2   The loss of groundwater

L1, L2   The groundwater recharge processing.

Figure 6.67 Results of independent inversion of the initial cross-hole ERT: a is the reference model, and the time-lapse inversion in percent changes at each subsequent monitoring time step ae depicted in b, c, d, and e (Liu *et al.*, 2017)[14]

to 22 Sept. Thus, the water inrush and discharge had serious impacts on the environment of groundwater. From 22 to 26 Sept., as shown in Figure 6.67(d), the value of resistivity decreased, and the groundwater level rose back to −76 m in the water-conducting structure (L1). The study also deduced that the curtain grouting in the tunnel face blocked the water, and the three consecutive days of rain aided in the gradual rising of the groundwater. On 30 Sept., the monitoring results showed that the groundwater level had arisen to −74 m in the water-conducting structure (L2), as shown in Figure 6.67, and water began to run again from the dried-up spring. It is probable that the vertical upward migration of the groundwater was the result of the previous rain and the replenishment of the upstream water. Based on the study, it was concluded that the TL-ERT method is sensitive to and effective in monitoring the migration of groundwater. The inversion images indicate clearly and directly the vertical migration of groundwater in the water-conducting structure.

6.9 Application of RI in cave detection

The ERI technique offers quick and cost-effective imaging of a subsurface resistivity pattern to a depth of several tens of meters. In combination with other geophysical techniques (e.g. ground penetration radar and seismic refraction), ERT provides a reliable model of subsurface structures and thus improves the interpretation of landform evolution. Pánek *et al.*, 2010

---

14 Reprinted from Tunnelling and Underground Space Technology, Vol 66, Bin Liu, Zhengyu Liu, Shucai Li, Kerui Fan, Lichao Nie, Xinxin Zhang, An improved Time-Lapse resistivity tomography to monitor and estimate the impact on the groundwater system induced by tunnel excavation, 107-120, Copyright (2017), with permission from Elsevier.

utilized the ERI techniques to detect crevice-type caves and other mass-movement-related discontinuities. The study also investigated the gravitational mass movement.

The Jaskinia Miecharska cave (MIC) is situated in the Beskid Śląski Mts. Its location is marked as "3" in Figure 6.68. According to Pánek *et al.*, 2010, the slope is formed by the Upper Godula beds of the Silesian Unit, where the beds are composed of thick bedded sandstones inter-bedded with shale, as presented in Figure 6.69. The Upper Godula beds belong to the southern flank of the Szczyrk Anticline. The rock strata strike N 260–290° and dip 12–20°, which makes them practically parallel to the dip of the slope. The rocks are cut by a joint system corresponding to regional master joint sets. The set of longitudinal joints (L) oriented ca. 260–280° and two sets of complementary diagonal joints (D1 = 130–140° and D2 = 220–240°) are well developed, while the set of transversal joints, T = ca. 340°, is less distinctly marked (Pánek *et al.*, 2010).

The MIC is an element of a large landslide covering above the valley bottom between two tributaries of Malinka creek, as presented in Figure 6.69A. The landslide represents a complex type of movement that combines translation (spreading), toppling, and rotation. These displacements were developed during several stages of mass movement, as presented in Figure 6.69A with L1 to L7. From the geomechanics point of view, the slope represents an intermediate and locally shallow type of gravitational slide (Pánek *et al.*, 2010).

According to Pánek *et al.*, 2010, the entrance of the cave is located in a trench at the foot of a secondary scarp of the transformed landslide. The MIC is 1810 m long, and its depth under the (inclined) ground is 10–20 m. The cave is composed of a maze of galleries and chambers which are 0.5–10 m high and 0.3–8 m wide and follow the orientation of the most prominent joint sets (diagonal D1 and D2 and transversal T). The cave system is divided into three parts, according to the location within the landslide and predominant direction of the cave galleries, as presented in Figure 6.69.

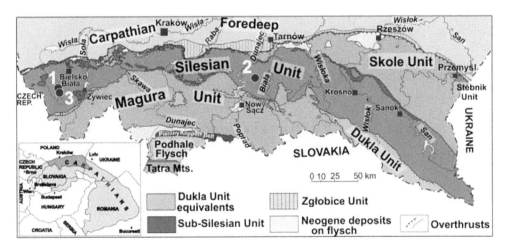

Figure 6.68 Geological map of the Polish Carpathians, showing the location of analyzed sites 3 -Jaskinia Miecharska Cave site (Pánek *et al.*, 2010)[15]

15 Reprinted from Geomorphology, Vol 123, Issue 1-2, Tomáš Pánek, Włodzimierz Margielewski, Petr Táboří̌k, Jan Urban, Jan Hradecký, Czesław Szura, Gravitationally induced caves and other discontinuities detected by 2D electrical resistivity tomography: Case studies from the Polish Flysch Carpathians,165-180, Copyright (2010), with permission from Elsevier.

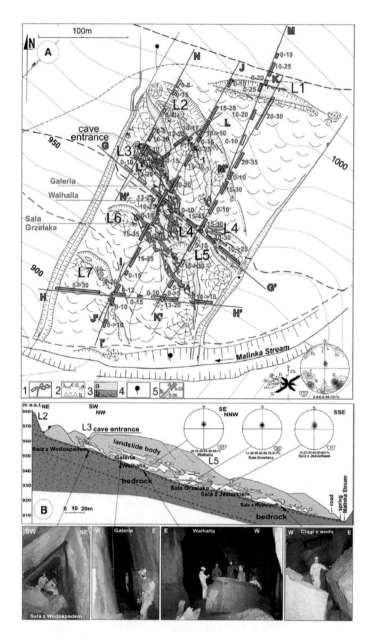

Figure 6.69 Jaskinia Miecharska Cave site: (a) Map of landslide and cave. L1 – L7 stages of land-slide development transforming the landslide zone; steeply dipping ERT profiles are shortened according to their projection on the horizontal plane; (b) cross section with the orientation of bedding planes (cave's bottom) in contour diagrams. Photos of the elements of the cave indicated with arrows (Pánek *et al.*, 2010)[16]

16 Reprinted from Geomorphology, Vol 123, Issue 1-2, Tomáš Pánek, Włodzimierz Margielewski, Petr Tábořík, Jan Urban, Jan Hradecký, Czesław Szura, Gravitationally induced caves and other discontinuities detected by 2D electrical resistivity tomography: Case studies from the Polish Flysch Carpathians,165-180, Copyright (2010), with permission from Elsevier.

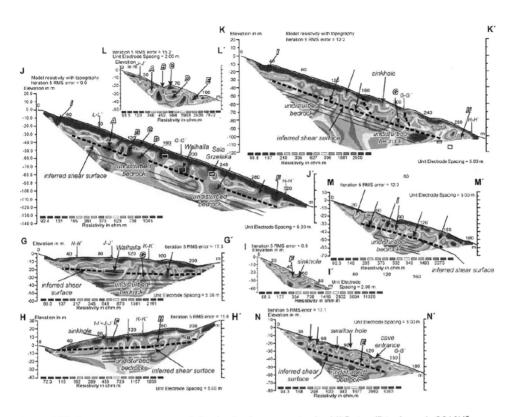

Figure 6.70 Resistivity tomograms of dipole-dipole surveys in the MIC site (Pánek et al., 2010)[17]

Pánek et al., 2010 surveyed eight ERI sections in the MIC site, including four long profiles with 5-m electrode spacing between two longitudinal and two transversal sections crossing the entire landslide body and four additional sections of shorter electrode spacing and higher resolution. The layout of the ERI test lines are presented in Figure 6.69. The 2D ERI test results of all of the sections are presented in Figure 6.70.

Based on the 2D ERI test results, Pánek et al., 2010 interpreted the high-resistivity anomaly as the concentration of fractures or other voids. The underlying conductive zone reflects the undisturbed bedrock, and the transition zone between both contrasting resistivity bodies can be interpreted as a slip surface (Figure 6.70). It should be noted that some of high-resistivity anomalies interrupt the underlying conductive zone in 2D ERI profiles, such as G – G', H – H', I – I' and J – J' profiles, which indicate fissure-related passages on the shear surface. Pánek et al., 2010 explored cave chambers with high-resistivity zones of N2000–3000 Ω m, such as the Walhalla chamber (15 m long, 8 m wide and 8 m high), Grzelak's chamber, and high Galeria passage situated in the vicinity of Walhalla (Figure 6.69). It was observed that

17 Reprinted from Geomorphology, Vol 123, Issue 1-2, Tomáš Pánek, Włodzimierz Margielewski, Petr Tábořík, Jan Urban, Jan Hradecký, Czesław Szura, Gravitationally induced caves and other discontinuities detected by 2D electrical resistivity tomography: Case studies from the Polish Flysch Carpathians,165-180, Copyright (2010), with permission from Elsevier.

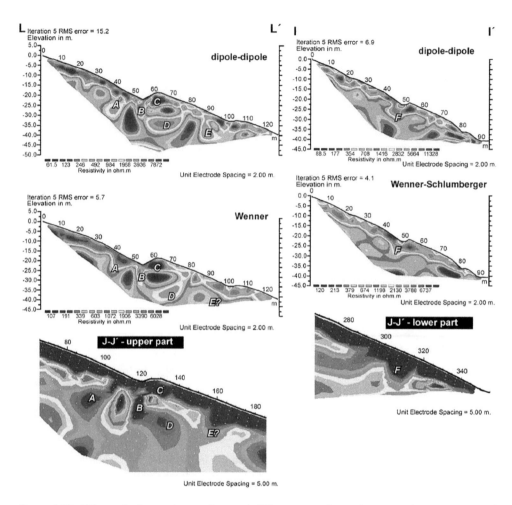

Figure 6.71 Effect of electrode spacing and different configurations on the geometry of discontinuities detected by selected profiles within the MIC site (Pánek *et al.*, 2010)[18]

these passages and chambers were well reflected, even with a 5 m electrode spacing (see G – G' and J – J' in Figure 6.70). The Walhalla chamber was easily detected in both longitudinal (J – J') and transversal (G – G') sections. Pánek *et al.*, 2010 emphasized that the location of this chamber, just on the slip surface, confirms the above interpretation of the transition zone between bodies of a contrasting resistivity pattern.

Pánek *et al.*, 2010 extended the study and investigated the application of various arrays and different electrode spacing to the L – L', I – I', and J – J' profiles. The 2D ERI test results are presented in Figure 6.71. It is noted that shorter electrode spacing assures higher resolution

18 Reprinted from Geomorphology, Vol 123, Issue 1-2, Tomáš Pánek, Włodzimierz Margielewski, Petr Táboŕík, Jan Urban, Jan Hradecký, Czesław Szura, Gravitationally induced caves and other discontinuities detected by 2D electrical resistivity tomography: Case studies from the Polish Flysch Carpathians,165-180, Copyright (2010), with permission from Elsevier.

of subsurface structures (e.g. the I – I', J – J', and L – L' profiles in the MIC; Figure 6.70). The ERI profiles studied using a 5 m electrode spacing displayed systems of crevices, whereas the sections with 2 m spacing revealed individual crevices. During the study, it was observed that when various electrode arrays are applied, shorter electrode spacing eliminates differences in the ERT record. (See the I – I' and L – L' profiles of the MIC site studied using a 2-m electrode spacing in Figure 6.71). From the viewpoint of the detection of crevices, differences between the dipole-dipole and the Wenner-Schlumberger or the Wenner arrays are rather negligible. The Wenner array along the L – L' section, as presented in Figure 6.71, displays the same structures (anomalies A, B, C, and D); however, the penetration is at slightly deeper levels. In addition, differences between the dipole-dipole and the Wenner-Schlumberger arrays are also small in the case of the I – I' profile in the MIC site.

Appendix A

Compacted clay correlations

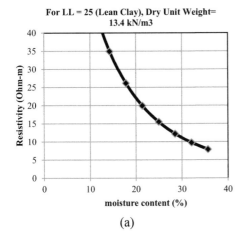

For LL = 25 (Lean Clay), Dry Unit Weight= 13.4 kN/m3

(a)

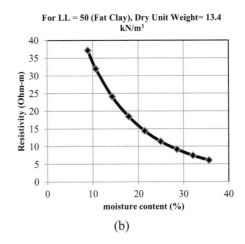

For LL = 50 (Fat Clay), Dry Unit Weight= 13.4 kN/m^3

(b)

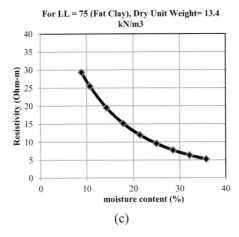

For LL = 75 (Fat Clay), Dry Unit Weight= 13.4 kN/m3

(c)

Figure A1 Evaluation of moisture content from resistivity for unit weight 13.4 kN/m^3

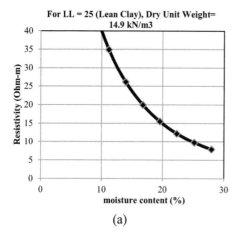

(a)

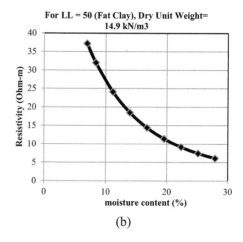

(b)

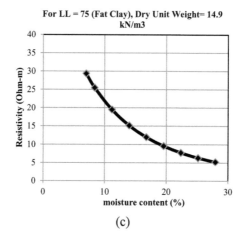

(c)

Figure A2 Evaluation of moisture content from resistivity for unit weight 14.9 kN/m³

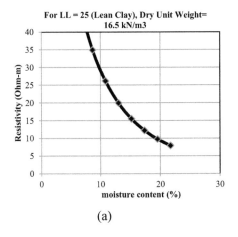

(a)

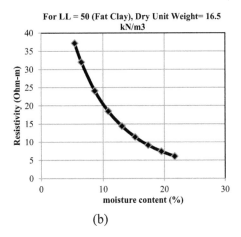

(b)

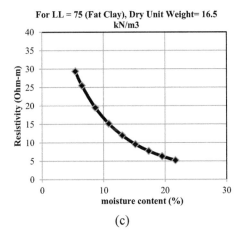

(c)

Figure A3 Evaluation of moisture content from resistivity for unit weight 16.5 kN/m³

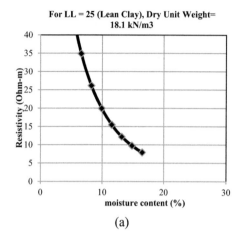

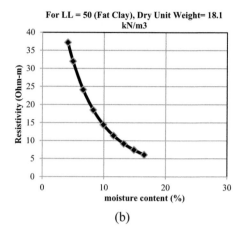

(a)

(b)

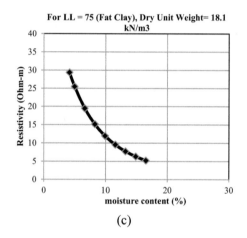

(c)

Figure A4 Evaluation of moisture content from resistivity for unit weight 16.5 kN/m³

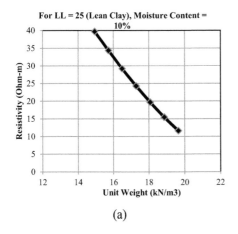

(a)

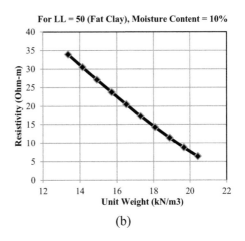

(b)

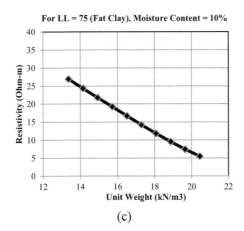

(c)

Figure A5 Evaluation of Unit Weight from resistivity for moisture content 10%

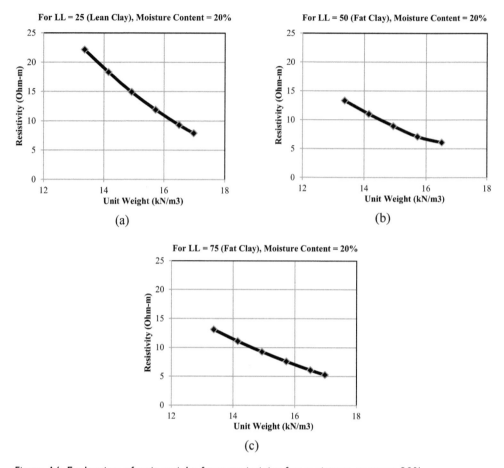

Figure A6 Evaluation of unit weight from resistivity for moisture content 20%

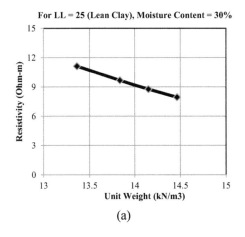

(a)

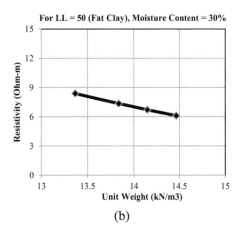

(b)

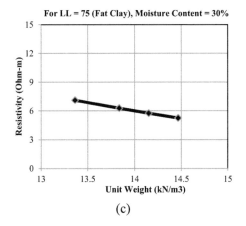

(c)

Figure A7 Evaluation of unit weight from resistivity for moisture content 30%

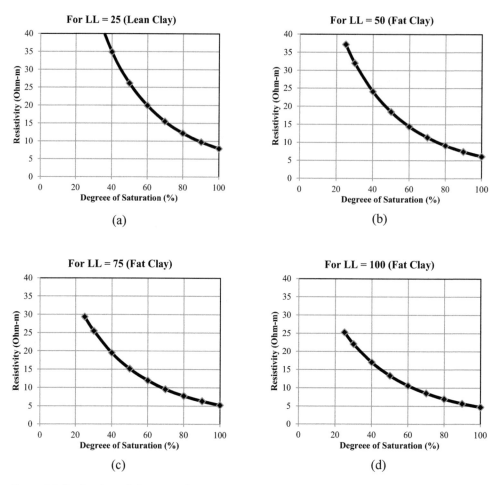

Figure A8 Evaluation of degrees of saturation using resistivity of compacted clay

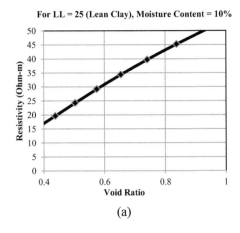

(a)

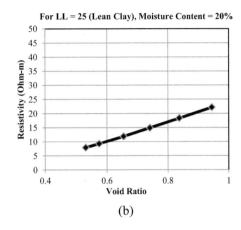

(b)

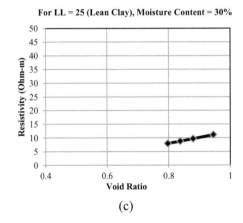

(c)

Figure A9 Evaluation of void ratio from resistivity for liquid limit = 25

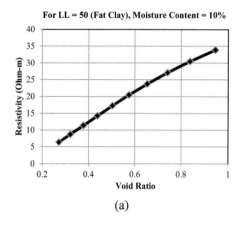

(a)

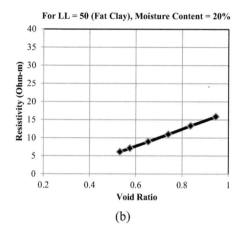

(b)

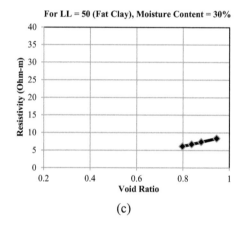

(c)

Figure A10 Evaluation of void ratio from resistivity for liquid limit = 50

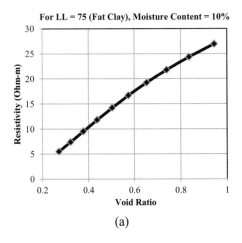

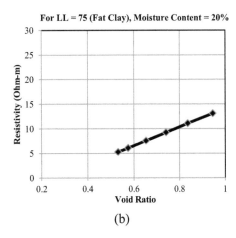

(a)

(b)

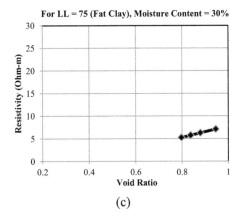

(c)

Figure A11 Evaluation of void ratio from resistivity for liquid limit = 75

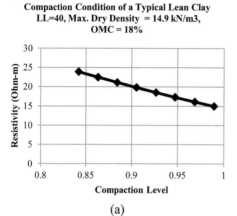

**Compaction Condition of a Typical Lean Clay
LL=40, Max. Dry Density = 14.9 kN/m3,
OMC = 18%**

(a)

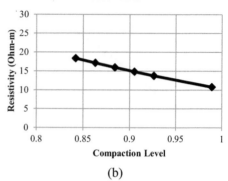

**Compaction Condition of a Typical Lean Clay
LL=40, Max. Dry Density = 14.9 kN/m3,
OMC = 22%**

(b)

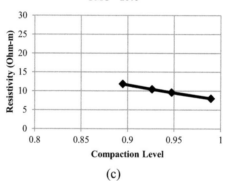

**Compaction Condition of a Typical Lean Clay
LL=40, Max. Dry Density = 14.9 kN/m3,
OMC = 26%**

(c)

Figure A12 Evaluation of compaction level from resistivity for liquid limit = 40, and Max. dry density = 14.9 kN/m³

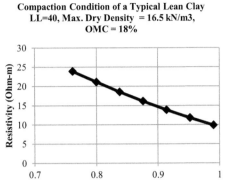

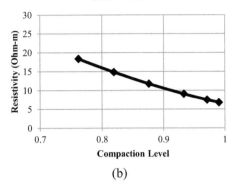

(a)

(b)

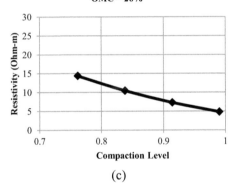

(c)

Figure A13 Evaluation of compaction level from resistivity for liquid limit = 40, and Max. dry density = 16.5 kN/m³

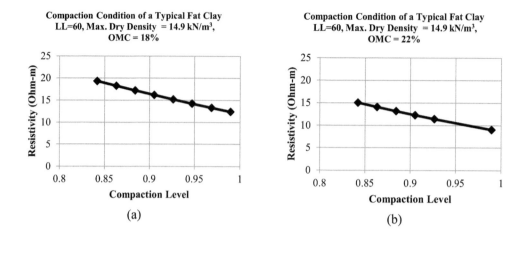

(a)

(b)

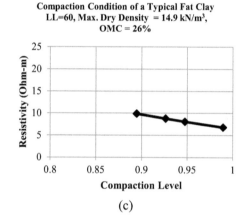

(c)

Figure A14 Evaluation of compaction level from resistivity for liquid limit = 60, and Max. dry density = 14.9 kN/m³

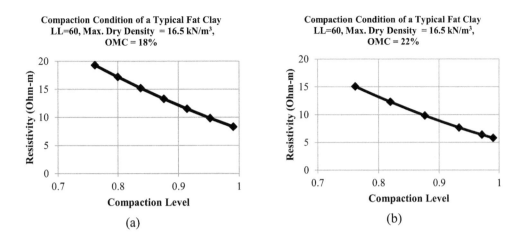

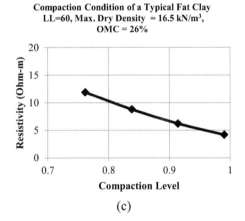

Figure A15 Evaluation of compaction level from resistivity for liquid limit = 60, and Max. dry density = 16.5 kN/m³

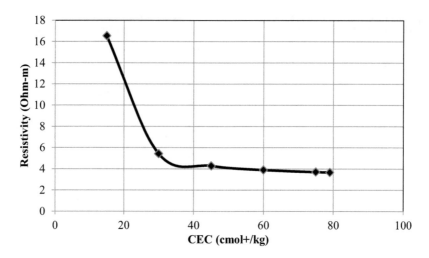

Figure A16 Evaluation of cation exchange capacity on resistivity

Appendix B

Undisturbed clay correlations

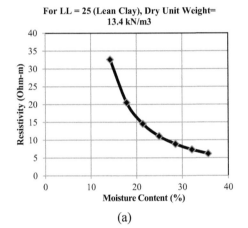

(a)

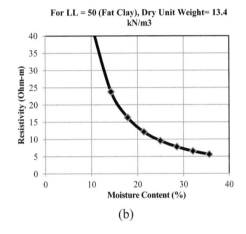

(b)

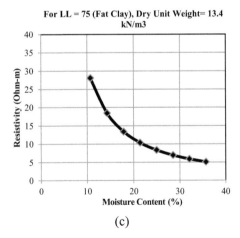

(c)

Figure B1 Evaluation of moisture content from resistivity for unit weight 13.4 kN/m³

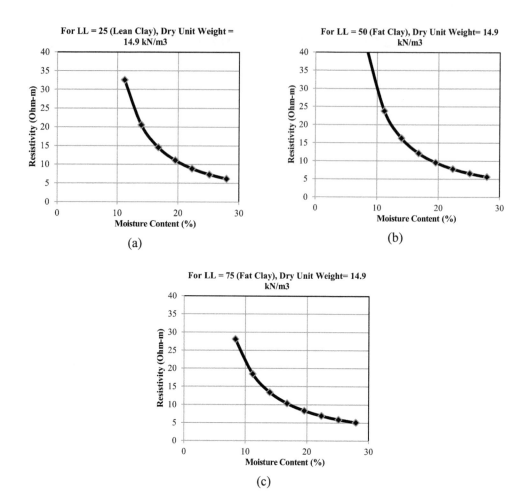

Figure B2 Evaluation of moisture content from resistivity for unit weight 14.9 kN/m³

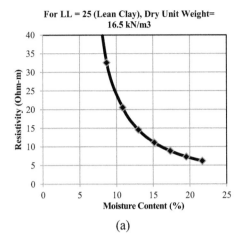

(a)

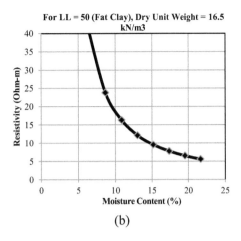

(b)

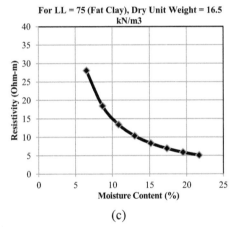

(c)

Figure B3 Evaluation of moisture content from resistivity for unit weight 16.5 kN/m³

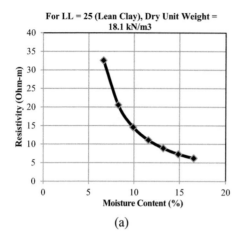

(a)

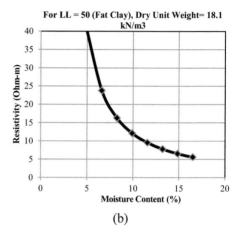

(b)

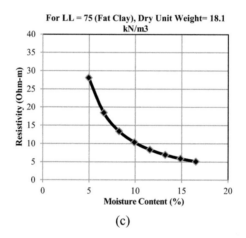

(c)

Figure B4 Evaluation of moisture content from resistivity for unit weight 18.1 kN/m³

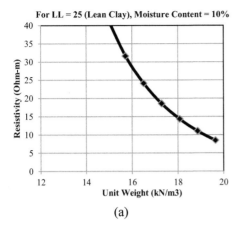

(a)

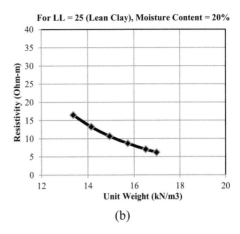

(b)

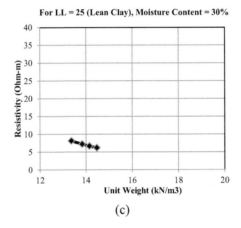

(c)

Figure B5 Evaluation of unit weight from resistivity for liquid limit = 25

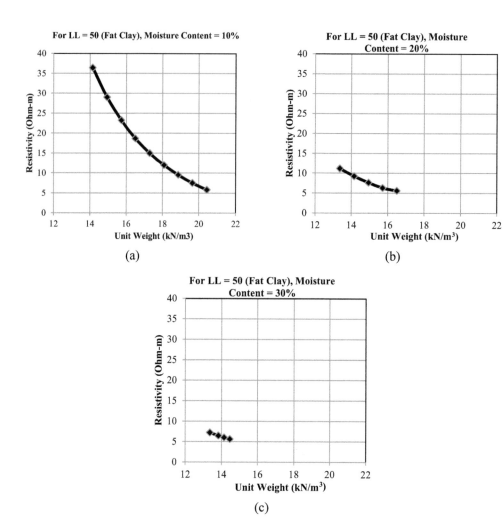

Figure B6 Evaluation of unit weight from resistivity for liquid limit = 50

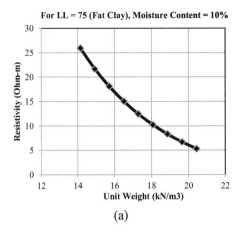

(a)

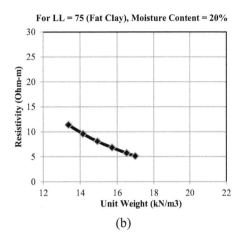

(b)

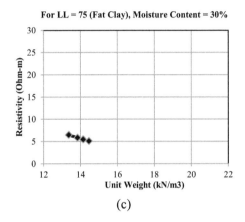

(c)

Figure B7 Evaluation of unit weight from resistivity for liquid limit = 75

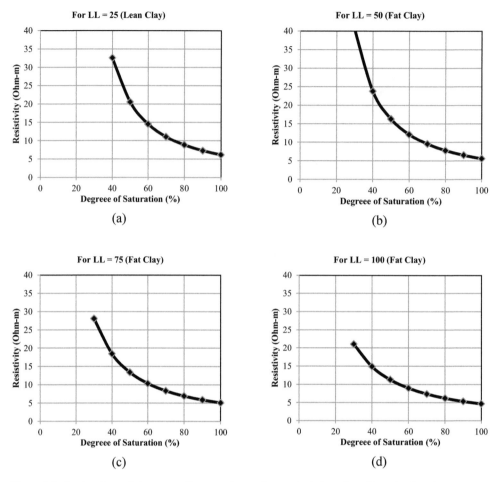

Figure B8 Evaluation of degrees of saturation using resistivity of undisturbed soils

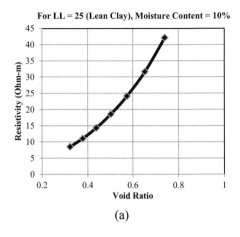

(a)

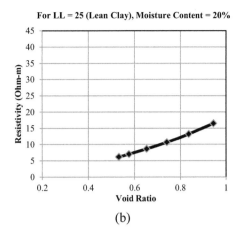

(b)

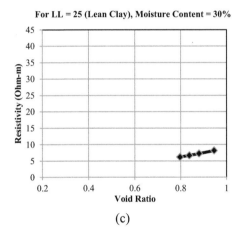

(c)

Figure B9 Evaluation of void ratio from resistivity for liquid limit = 25

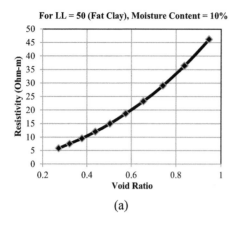

(a)

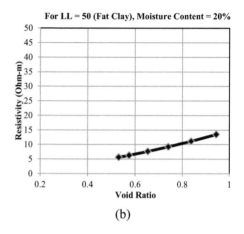

(b)

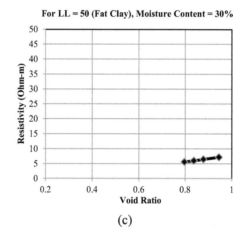

(c)

Figure B10 Evaluation of void ratio from resistivity for liquid limit = 50

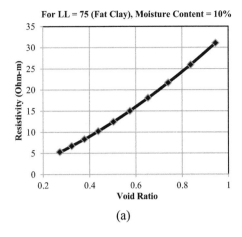

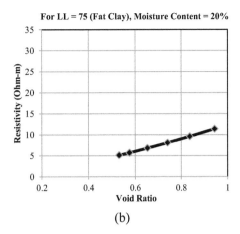

(a)

(b)

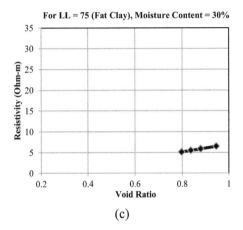

(c)

Figure B11 Evaluation of void ratio from resistivity for liquid limit = 75

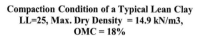

Compaction Condition of a Typical Lean Clay
LL=25, Max. Dry Density = 14.9 kN/m3,
OMC = 18%

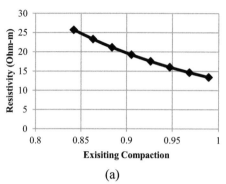

(a)

Compaction Condition of a Typical Lean Clay
LL=25, Max. Dry Density = 14.9 kN/m3,
OMC = 22%

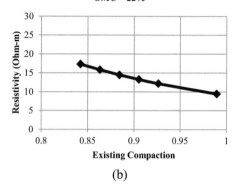

(b)

Compaction Condition of a Typical Lean Clay
LL=25, Max. Dry Density = 14.9 kN/m3,
OMC = 26%

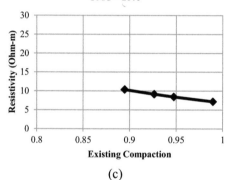

(c)

Figure B12 Evaluation of existing compaction from resistivity for liquid limit = 25, Max. dry density = 14.9 kN/m³

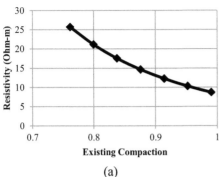

Compaction Condition of a Typical Lean Clay
LL=25, Max. Dry Density = 16.5 kN/m3,
OMC = 18%

(a)

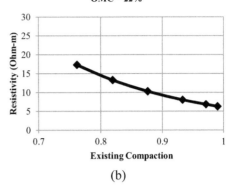

Compaction Condition of a Typical Lean Clay
LL=25, Max. Dry Density = 16.5 kN/m3,
OMC = 22%

(b)

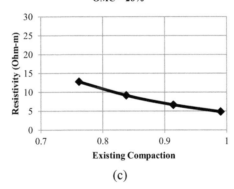

Compaction Condition of a Typical Lean Clay
LL=25, Max. Dry Density = 16.5 kN/m3,
OMC = 26%

(c)

Figure B13 Evaluation of existing compaction from resistivity for liquid limit = 25, Max. dry density = 16.5 kN/m³

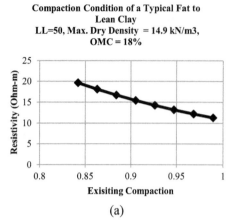

(a)

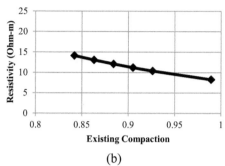

(b)

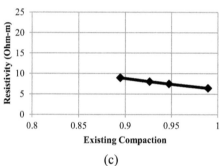

(c)

Figure B14 Evaluation of existing compaction from resistivity for liquid limit = 50, Max. dry density = 14.9 kN/m³

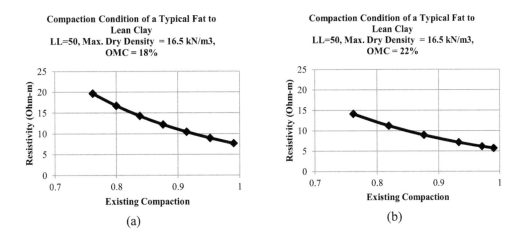

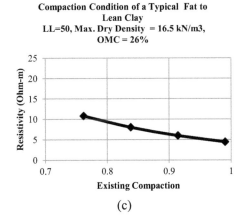

Figure B15 Evaluation of existing compaction from resistivity for liquid limit = 25, Max. dry density = 16.5 kN/m³

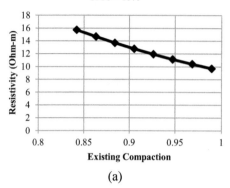

(a)

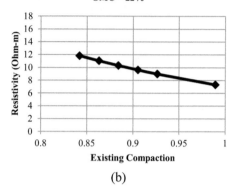

(b)

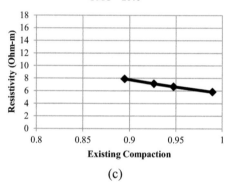

(c)

Figure B16 Evaluation of existing compaction from resistivity for liquid limit = 75, Max. dry density = 14.9 kN/m³

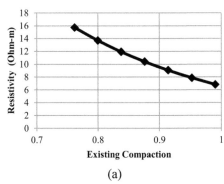

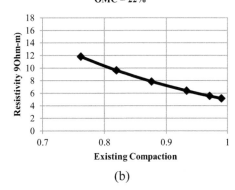

(a)

(b)

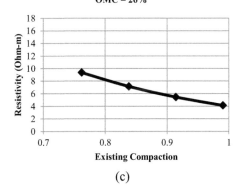

(c)

Figure B17 Evaluation of existing compaction from resistivity for liquid limit=75, Max. dry density = 16.5 kN/m³

Appendix C

Boring logs

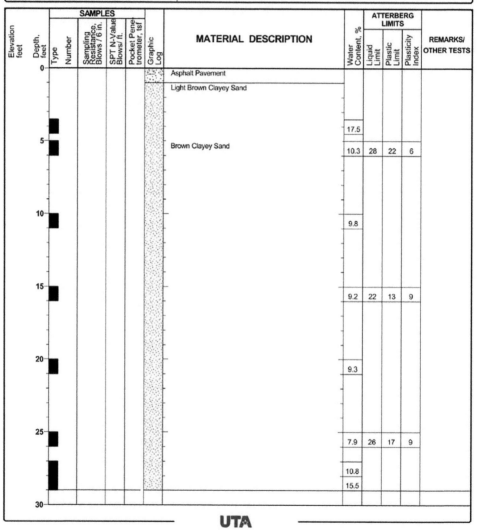

Project: Failure Analysis of MSE wall	Log of Boring BH-1
Project Location: SH342, Lancaster, TX.	
Project Number:	Sheet 1 of 1

Date(s) Drilled	10/15/09 - 10/15/09	Logged By	UTA	Checked By	UTA
Drilling Method		Drill Bit Size/Type		Total Depth of Borehole	29.0
Drill Rig Type	Hollow stem	Drilling Contractor		Surface Elevation	
Groundwater Level(s)		Sampling Method(s)	Disturbed Sample	Hammer Data	
Borehole Backfill	Inclinometer casing installed at the bore hole	Comments			

Figure C1 Borehole log of boring 1 at the MSE wall

UTA

Material Descriptions in graphic log:
- Asphalt Pavement
- Light Brown Clayey Sand
- Brown Clayey Sand

Water Content and Atterberg values at depths:
- ~4 ft: 17.5
- ~5 ft: 10.3, Liquid Limit 28, Plastic Limit 22, Plasticity Index 6
- ~10 ft: 9.8
- ~15 ft: 9.2, Liquid Limit 22, Plastic Limit 13, Plasticity Index 9
- ~21 ft: 9.3
- ~25 ft: 7.9, Liquid Limit 26, Plastic Limit 17, Plasticity Index 9
- ~27 ft: 10.8
- ~28 ft: 15.5

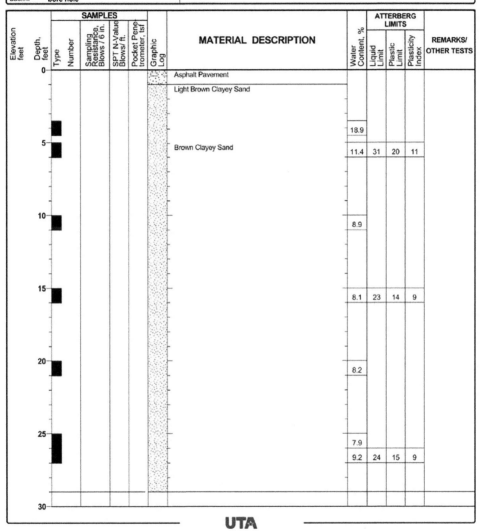

Project: Failure Analysis of MSE wall	Log of Boring BH-2
Project Location: SH342, Lancaster, TX.	Sheet 1 of 1
Project Number:	

Date(s) Drilled	10/15/09 - 10/15/09	Logged By	UTA	Checked By	UTA
Drilling Method		Drill Bit Size/Type		Total Depth of Borehole	29.0
Drill Rig Type	Hollow stem	Drilling Contractor		Surface Elevation	
Groundwater Level(s)		Sampling Method(s)	Disturbed Sample	Hammer Data	
Borehole Backfill	Inclinometer casing installed at the bore hole	Comments			

MATERIAL DESCRIPTION

- Asphalt Pavement
- Light Brown Clayey Sand
- Brown Clayey Sand

Water Content, % / ATTERBERG LIMITS (Liquid Limit, Plastic Limit, Plasticity Index)

Depth	Water Content, %	Liquid Limit	Plastic Limit	Plasticity Index
~4	18.9			
5	11.4	31	20	11
10	8.9			
15	8.1	23	14	9
20	8.2			
25	7.9			
~26	9.2	24	15	9

UTA

Figure C2 Borehole log of boring 2 at the MSE wall

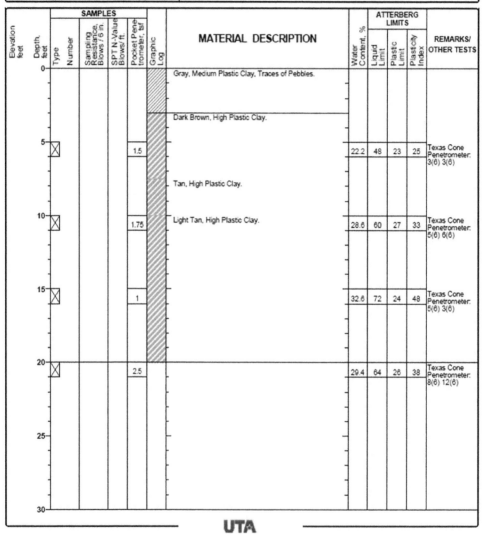

Project: Slope Stabiliy Analysis	Log of Boring BH-1
Project Location: US 287 (North), Mansfield, TX.	Sheet 1 of 1
Project Number:	

Date(s) Drilled	10/28/10 - 10/28/10	Logged By	UTA	Checked By	UTA
Drilling Method		Drill Bit Size/Type		Total Depth of Borehole	20.0
Drill Rig Type	Hollow stem	Drilling Contractor	Alpha Testing, Inc.	Surface Elevation	
Groundwater Level(s)		Sampling Method(s)	Shelby Tube	Hammer Data	
Borehole Backfill	Moisture Sensor installed at the bore hole.	Comments			

SAMPLES / **MATERIAL DESCRIPTION** / **ATTERBERG LIMITS** / **REMARKS/OTHER TESTS**

Columns: Elevation feet | Depth, feet | Type | Number | Sampling Resistance, Blows / 6 in. | SPT N-Value Blows/ ft | Pocket Penetrometer, tsf | Graphic Log | Water Content, % | Liquid Limit | Plastic Limit | Plasticity Index

0 — Gray, Medium Plastic Clay, Traces of Pebbles.

Dark Brown, High Plastic Clay.

5 — | 1.5 | | | Tan, High Plastic Clay. | 22.2 | 48 | 23 | 25 | Texas Cone Penetrometer: 3(6) 3(6)

10 — | 1.75 | Light Tan, High Plastic Clay. | 28.6 | 60 | 27 | 33 | Texas Cone Penetrometer: 5(6) 6(6)

15 — | 1 | | 32.6 | 72 | 24 | 48 | Texas Cone Penetrometer: 5(6) 3(6)

20 — | 2.5 | | 29.4 | 64 | 26 | 38 | Texas Cone Penetrometer: 8(6) 12(6)

25 —

30 —

UTA

Figure C3 Log of BH-1 at US 287 slope

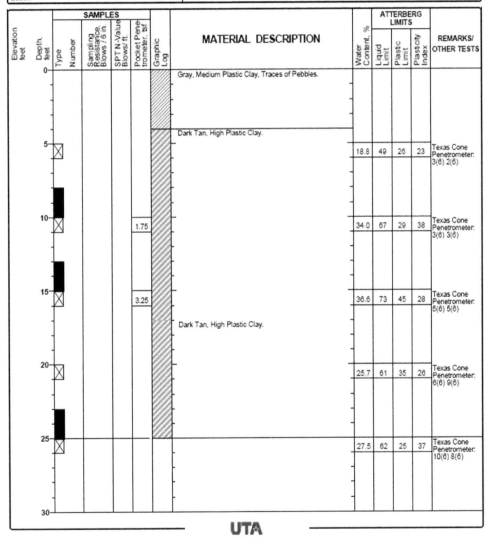

		SAMPLES				MATERIAL DESCRIPTION		ATTERBERG LIMITS				

Project: Slope Stabiliy Analysis
Project Location: US 287 (North), Mansfield, TX.
Project Number:

Log of Boring BH-2
Sheet 1 of 1

Date(s) Drilled	10/28/10 - 10/28/10	Logged By	UTA	Checked By	UTA
Drilling Method		Drill Bit Size/Type		Total Depth of Borehole	25.0
Drill Rig Type	Hollow stem	Drilling Contractor	Alpha Testing, Inc.	Surface Elevation	
Groundwater Level(s)		Sampling Method(s)	Shelby Tube	Hammer Data	
Borehole Backfill	Cement Grout	Comments			

Material Description column:
- 0': Gray, Medium Plastic Clay, Traces of Pebbles.
- ~4': Dark Tan, High Plastic Clay.
- ~17': Dark Tan, High Plastic Clay.

Pocket Penetrometer / SPT data:
- Depth 10: 1.75
- Depth 15: 3.25

Atterberg Limits / Water Content / Remarks:

Depth	Water Content, %	Liquid Limit	Plastic Limit	Plasticity Index	REMARKS/OTHER TESTS
5	18.8	49	26	23	Texas Cone Penetrometer: 3(6) 2(6)
10	34.0	67	29	38	Texas Cone Penetrometer: 3(6) 3(6)
15	36.6	73	45	28	Texas Cone Penetrometer: 5(6) 5(6)
20	25.7	61	35	26	Texas Cone Penetrometer: 6(6) 9(6)
25	27.5	62	25	37	Texas Cone Penetrometer: 10(6) 8(6)

UTA

Figure C4 Log for borehole BH-2 at US 287 slope

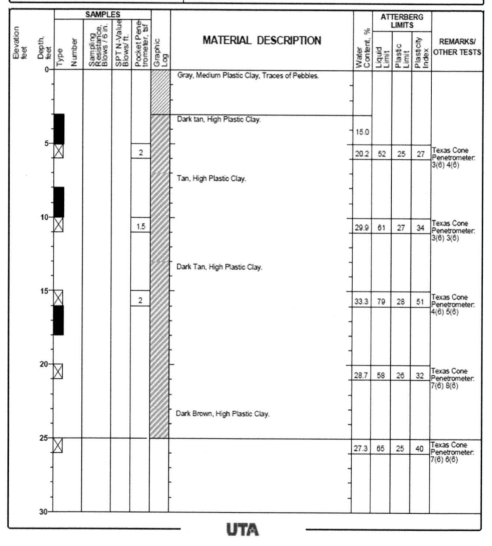

Date(s) Drilled	10/28/10 - 10/28/10	Logged By	UTA	Checked By	UTA
Drilling Method		Drill Bit Size/Type		Total Depth of Borehole	25.0
Drill Rig Type	Hollow stem	Drilling Contractor	Alpha Testing, Inc.	Surface Elevation	
Groundwater Level(s)		Sampling Method(s)	Shelby Tube	Hammer Data	
Borehole Backfill	Cement Grout	Comments			

MATERIAL DESCRIPTION

Gray, Medium Plastic Clay, Traces of Pebbles.

Dark tan, High Plastic Clay.

Tan, High Plastic Clay.

Dark Tan, High Plastic Clay.

Dark Brown, High Plastic Clay.

Depth, feet	Number	Pocket Penetrometer, tsf	Water Content, %	Liquid Limit	Plastic Limit	Plasticity Index	REMARKS/OTHER TESTS
			15.0				
5	2		20.2	52	25	27	Texas Cone Penetrometer: 3(6) 4(6)
10	1.5		29.9	61	27	34	Texas Cone Penetrometer: 3(6) 3(6)
15	2		33.3	79	28	51	Texas Cone Penetrometer: 4(6) 5(6)
20			28.7	58	26	32	Texas Cone Penetrometer: 7(6) 8(6)
25			27.3	65	25	40	Texas Cone Penetrometer: 7(6) 6(6)

UTA

Figure C5 Log for borehole BH-3 at US 287 slope

References

Abu-Hassanein, Z.S., Benson, C.H. & Blotz, L.R. (1996) Electrical resistivity of compacted clays. *Journal of Geotechnical and Geoenvironmental Engineering*, 122(5), 397–406.

Abu-Hassanein, Z.S., Benson, C.H., Wang, X. & Blotz, L.R. (1996) Determining bentonite content in soil-bentonite mixtures using electrical conductivity. *Geotechnical Testing Journal*, 19(1), 51–57.

Abu-Hassanein, Z.S., Benson, C.H., & Blotz, L.R. (1996). Electrical resistivity of compacted clays. *Journal of Geotechnical Engineering*, 122(5), 397–406.

Abrams, T.G., Wright, S.G., 1972. A survey of earth slope failures and remedial measures in Texas (Research Report 161–1). Center for Transportation Research. University of Texas at Austin, Austin, TX.

Advanced Geosciences, Inc. (2006) Instruction Manual for EarthImager 2D Version 1.7.4. Resistivity and IP Inversion Software, Austin, Texas. Available from www.agiusa.com.

Aizebeokhai, A.P. (2010) 2D and 3D geoelectrical resistivity imaging: Theory and field design. *Scientific Research and Essays*, 5(23), 3592–3605.

Alam, M.Z., Hossain, M.S. & Samir, S. (2017) Performance evaluation of a bioreactor landfill operation. In: *Geotechnical Frontiers*. pp.267–273.

American Water Works Association. (1999) ANSI standard for poly-ethylene encasement for ductile-iron pipe system. C105/A21.5–99, Denver.

Archie, G.E. (1942) The electrical resistivity log as an aid in determining some reservoir characteristics. *Petroleum Technology, T.P. 1422*, Shell Oil Co. Houston, TX.

Arjwech, R. (2011) *Electrical Resistivity Imaging for Unknown Bridge Foundation Depth Determination*. Ph.D. Dissertation, Texas A & M University, College Station, TX.

Arulanandan, K. (1969) Hydraulic and electrical flows in clays. *Clays Clay Minerals*, 17, 63–76.

ASTM D1586-11, Standard Test Method for Standard Penetration Test (SPT) and Split-Barrel Sampling of Soils, ASTM International, West Conshohocken, PA, 2011, www.astm.org.

ASTM D4633-16, Standard Test Method for Energy Measurement for Dynamic Penetrometers, ASTM International, West Conshohocken, PA, 2016, www.astm.org.

ASTM D3441-16, Standard Test Method for Mechanical Cone Penetration Testing of Soils, ASTM International, West Conshohocken, PA, 2016, www.astm.org.

ASTM D5778-12, Standard Test Method for Electronic Friction Cone and Piezocone Penetration Testing of Soils, ASTM International, West Conshohocken, PA, 2012, www.astm.org.

ASTM D4719-07, Standard Test Methods for Prebored Pressuremeter Testing in Soils (Withdrawn 2016), ASTM International, West Conshohocken, PA, 2007, www.astm.org.

ASTM D6635-15, Standard Test Method for Performing the Flat Plate Dilatometer, ASTM International, West Conshohocken, PA, 2015, www.astm.org.

ASTM D2573 / D2573M-15e1, Standard Test Method for Field Vane Shear Test in Saturated Fine-Grained Soils, ASTM International, West Conshohocken, PA, 2015, www.astm.org.

ASTM G187-12a, Standard Test Method for Measurement of Soil Resistivity Using the Two-Electrode Soil Box Method, ASTM International, West Conshohocken, PA, 2012, www.astm.org.

ASTM D4318-17, Standard Test Methods for Liquid Limit, Plastic Limit, and Plasticity Index of Soils, ASTM International, West Conshohocken, PA, 2017, www.astm.org.

Barlaz, M.A., Ham, R.K. & Schaefer, D.M. (1990) Methane production from municipal refuse: A review of enhancement techniques and microbial dynamics. *Critical Reviews in Environmental Control*, 19(6), 557–584.

Besson, A., Cousin, I., Dorigny, A., Dabas, M. & King, D. (2008) The temperature correction for the electrical resistivity measurements in undisturbed soil samples: Analysis of the existing conversion models and proposal of a new model. *Soil Science*, 173(10), 707–720.

Brunet, P., Clément, R. & Bouvier, C. (2010) Monitoring soil water content and deficit using electrical resistivity tomography (ERT) – A case study in the Cevennes area, France. *Journal of Hydrology*, 380(1–2), 146–153.

Bryson, L.S. (2005) Evaluation of geotechnical parameters using electrical resistivity measurements. *Proceedings, Earthquake Engineering and Soil Dynamics, GSP 133, Geo-Frontiers 2005*, ASCE, Reston, VA.

Caltrans. (2003) *Corrosion Guidelines*. Version 1. Sacramento, CA.

Campbell, R.B., Bower, C.A. & Richards, L.A. (1948) Change of electrical conductivity with temperature and the relation of osmotic pressure to electrical conductivity and ion concentration for soil extracts. *Soil Science Society of America Journal*, 13, 66–69.

Clement, R., Descloitres, M., Gunther, T., Oxarango, L., Morra, C., Laurent, J., *et al.* (2010) Improvement of electrical resistivity tomography for leachate injection monitoring. *Waste Management*, 30, 452–464.

Comina, C., Foti, S., Musso, G. & Romero, E. (2008) EIT Oedometer: An advanced cell to monitor spatial and time variability in soil with electrical and seismic measurements. *Geotechnical Testing Journal*, 31(5), 564.

Crony, D., Coleman, J. & Currer, E. (1951) The electrical resistance method of measuring soil moisture. *British Journal of Applied Physics*, 2(4), 85–91.

Dahlin, T. (2001) The development of DC resistivity imaging techniques. *Computers & Geosciences*, 27(9), 1019–1029.

Daniel, D.E & Benson, C.H. (1990) Water content-density criteria for compacted soil liners. *Journal of Geotechnical Engineering*, 116(12), 1811–1830.

Dixon, N. & Jones, D.R.V. (2005) Engineering properties of municipal solid waste. *Geotextiles and Geomembranes*, 23(3), 205–233.

Ekwue, E. & Bartholomew, J. (2010) Electrical conductivity of some soils in Trinidad as affected by density, water and peat content. *Biosystems Engineering*, 108(2), 95–103.

EPA. (2009) Municipal solid waste generation, recycling, and disposal in the United States: Fact and figures for 2009.

Farrar, D. & Coleman, J. (1967) The correlation of surface area with other properties of nineteen British clay soils. *Journal of Soil Science*, 18, 118–124.

Friedman, S.P. (2005) Soil properties influencing apparent electrical conductivity: A review. *Computers and Electronics in Agriculture*, 46(1), 45–70.

Fujimoto, K. (2009) Application of the resistivity imaging method to identify seepage flow paths. *Case Study: Lewisville Dam, Lewisville, TX, Master of Engineering in Civil Engineering Project*, University of Texas at Arlington, TX.

Fukue, M., Minato, T., Horibe, H. & Taya, N. (1999) The micro-structures of clay given by resistivity measurements. *Engineering Geology*, 54(1–2), 43–53.

Giao, P., Chung, S., Kim, D. & Tanaka, H. (2003) Electric imaging and laboratory resistivity testing for geotechnical investigation of Pusan clay deposits. *Journal of Applied Geophysics*, 52(4), 157–175.

Goyal, V.C., Gupta, P.K., Seth, S.M. & Singh, V.N. (1996) Estimation of temporal changes in soil moisture using resistivity method. *Journal of Hydrological Processes*, 10, 1147–1154.

Gassman, S.L., & Finno, R.J. (1999) Impulse response evaluation of foundations using multiple geophones. *Journal of Performance of Constructed Facilities*, 13(2), 82–89.

Grellier, S., Bouye, J., Guerin, R., Robain, H. & Skhiri, N. (2005) Electrical resistivity tomography (ERT) applied to moisture measurements in bioreactor: Principles, *in-situ* measurements and results. *Proceedings, International Workshop Hydro-Physico-Mechanics of Landfills*, Grenoble, France.

Grellier, S., Reddy, K.R., Gangathulasi, J., Adib, R. & Peters, C.C. (2007) Correlation between electrical resistivity and moisture content of municipal solid waste in bioreactor landfill. *Geotechnical Special Publication*, 163.

Grellier, S., Robain, H., Bellier, G. & Skhiri, N. (2006) Influence of temperature on the electrical conductivity of leachate from municipal solid waste. *Journal of Hazardous Materials*, 612–617.

Guerin, R., *et al.* (2004) Leachate recirculation: Moisture content assessment by means of a geophysical technique. *Waste Management*, 24(8), 785–794.

Hakonson, T.E. (1997) Capping as an alternative for landfill closures-perspectives and approaches. *Environmental Science and Research Foundation, Proceedings, Landfill Capping in the Semi Arid West: Problems, Perspectives, and Solutions*. Grand Teton National Park, Wyoming, May 21–22, ESRF,1–18.

Hanson, J., Yesiller, N. & Oettle, N. (2010) Spatial and temporal temperature distributions in municipal solid waste landfills. *Journal of Environmental Engineering*, 136(8), 804–814.

Haque, A. (2007) *Dynamic Characteristics and Stability Analysis of MSW in Bioreactor Landfills*. Ph.D. Dissertation, University of Texas at Arlington, Arlington, TX.

Holeman, J.N. (1970) Clay minerals. *Soil Conservation Service Engineering Division, US Department of Agriculture*, Washington, DC, USA, Technical Release No. 28.

Holtz, R.D. & Kovacs, W.D. (1981) *An Introduction to Geotechnical Engineering*. Prentice-Hall, Inc., Englewood Cliffs, NJ.

Hossain, J. (2012) *Geohazard Potential of Rainfall Induced Slope Failure on Expansive Clay*. Ph.D. Dissertation, University of Texas at Arlington, TX.

Hossain, M.S., Dharmateja, M. & Hossain, J. (2010) Assessment of geo-hazard potential and site investigations using resistivity imaging. *International Journal of Environmental Technology and Management*, 13(2), 116–129.

Hossain, M.S., Khan, M.S., Hossain, J., Kibria, G. & Taufiq, T. (2011) Evaluation of unknown foundation depth using different NDT methods. *Journal of Performance of Constructed Facilities*, 27(2), 209–214.

Hossain, M.S., Kibria, G., Khan, M.S., Hossain, J. & Taufiq, T. (2012) Effects of backfill soil on excessive movement of MSE wall. *Journal of Performance of Constructed Facilities*, 26(6), 793–802.

Hossain, M.S., Kibria, G. & Lozano, N. (2014) Determination of the depth and discontinuity of hard limestone stratum using resistivity imaging (RI). In: *Geo-Congress 2014: Geo-characterization and Modeling for Sustainability*. pp.2394–2403.

Hubbard, J.L. (2010) *Use of Electrical Resistivity and Multichannel Analysis of Surface Wave Geophysical Tomography in Geotechnical Site Characterization of Dam*. M.S. Thesis, The University of Texas at Arlington, Arlington, TX.

Hsu, S.C., Nelson, P.P. (2002) Characterization of eagle ford clay engineering. *Enginerring Geology*, 67(1–2), 169–183.

Imhoff, P.T., Reinhart, D.R., Englund, M., Guerin, R., Gawande, N., Han, B., *et al.* (2007) Review of state of the art methods for measuring water in landfills. *Waste Management*, 27, 729–745.

Jackson, P.D., Smith, D.T. & Stanford, P.N. (1978) Resistivity porosity-particle shape relationships for marine sand. *Geophysics*, 43(6), 1250–1268.

Jakosky, J.J. (1950) *Exploration Geophysics*. Trija Pub. Co., Los Angeles, 119.

Kalinski, R. & Kelly, W. (1993) Estimating water content of soils from electrical resistivity. *Geotechnical Testing Journal. ASTM*, 16(3), 323–329.

Keller, G. & Frischnecht, F. (1966) *Electrical Methods in Geophysical Prospecting*. Pergamon Press, New York.

Khan, M.S., Hossain, M.S., Hossain, J. & Kibria, G. (2012) Determining unknown bridge foundation depth by resistivity imaging method. In *GeoCongress 2012: State of the Art and Practice in Geotechnical Engineering* (275–284).

Khan, M.S., Hossain, S., & Kibria, G. (2015). Slope stabilization using recycled plastic pins. *Journal of Performance of Constructed Facilities*, 30(3), 04015054.

Khan, M.S., Hossain, S., Ahmed, A. & Faysal, M. (2017) Investigation of a shallow slope failure on expansive clay in Texas. *Engineering Geology*, 219, 118–129.

Kibria, G. (2011) *Determination of Geotechnical Properties of Clayey Soils From Resistivity Imaging (RI)*. MS Thesis, University of Texas at Arlington.

Kibria, G. (2014) *Evaluation of Physico-mechanical Properties of Clayey Soils Using Electrical Resistivity Imaging Technique*. Ph.D. Dissertation, University of Texas at Arlington.

Kibria, G. & Hossain, M.S. (2012) Investigation of geotechnical parameters affecting electrical resistivity of compacted clays. *Journal of Geotechnical and Geoenvironmental Engineering*, 138(12), 1520–1529.

Kibria, G. & Hossain, M.S. (2014) Effects of bentonite content on electrical resistivity of soils. In: *Geo-Congress 2014: Geo-characterization and Modeling for Sustainability*. pp.2404–2413.

Kibria, G. & Hossain, M.S. (2015) Investigation of degree of saturation in landfill liners using electrical resistivity imaging. *Waste Management*, 39, 197–204.

Kibria, G. & Hossain, M.S. (2016) Quantification of degree of saturation at shallow depths of earth slopes using resistivity imaging technique. *Journal of Geotechnical and Geoenvironmental Engineering*, 142(7), 06016004.

Kibria, G. & Hossain, M.S. (2017) Electrical resistivity of compacted clay minerals. *Environmental Geotechnics*. Available from: https://doi.org/10.1680/jenge.16.00005, ICE.

Kibria, G. & Hossain, M.S. (2017) Evaluation of corrosion potential of subsoil using geotechnical properties. *Journal of Pipeline Systems Engineering and Practice*, 9(1), 04017033.

Kim, J.H., Yoon, H.K. & Lee, J.S. (2011) Void ratio estimation of soft soils using electrical resistivity cone probe. *Journal of Geotechnical and Geoenvironmental Engineering*, 137(1), 86–93.

Kutner, M.H., Nachtscheim, C.J., Neter, J. & Li, W. (2005) *Applied Linear Statistical Model*. 5th ed., McGraw Hill Inc., New York.

Lake Gladewater Dam Inspection Report. *Texas Commission of Environmental Quality. (2005) Inspection Report*, Austin, TX, 2005.

Liu, B., Liu, Z., Li, S., Fan, K., Nie, L. & Zhang, X. (2017) An improved time-lapse resistivity tomography to monitor and estimate the impact on the groundwater system induced by tunnel excavation. *Tunneling and Underground Space Technology*, 66, 107–120.

Liu, W. (2007) *Thermal Analysis of Landfills*. PhD Dissertation, Wayne State University, Michigan.

Loke, M. (1999) A practical guide to 2D and 3D surveys. *Electrical Imaging Surveys for Environmental and Engineering Studies*, Minden Heights, 11700 Penang, Malaysia.

Loke, M.H. (2000) Electrical imaging surveys for environmental and engineering studies. *A Practical Guide to 2-D and 3-D Surveys*.

Manzur, S.R. (2013) *Hydraulic performance evaluation of different recirculation systems for ELR/bioreactor landfills*. Ph.D. Dissertation, University of Texas at Arlington, TX.

Manzur, S.R., Hossain, M.S., Kemler, V. & Khan, M.S. (2016) Monitoring extent of moisture variations due to leachate recirculation in an ELR/bioreactor landfill using resistivity imaging. *Waste Management*, 55, 38–48.

Matsui, T., Park, S.G., Park, M.K. & Matsuura, S. (2000, November) Relationship between electrical resistivity and physical properties of rocks. *Proceedings of an International Conference on Geotechnical & Geological Engineering, "GeoEng2000"*, Melbourne, Australia G (Vol. 987).

McCarter, W. (1984) The electrical resistivity characteristics of compacted clays. *Geotechnique*, 34(2), 263.

McCarter, W.J., Blewett, J., Chrisp, T.M. & Starrs, G. (2005) Electrical property measurements using a modified hydraulic oedometer. *Canadian Geotechnical Journal*, 42(2), 655–662.

McCarter, W.J. & Desmazes, P.(1997) Soil characterization using electrical measurements. *Geotechnique*, 47(1), 179–183.

Mitchell, J. & Soga, K. (2005) *Fundamentals of Soil Behavior*. John Wiley and Sons, Inc., Hoboken, NJ.

Mojid, M.A. & Cho, H. (2006) Estimating the fully developed diffuse double layer thickness from the bulk electrical conductivity in clay. *Applied Clay Science*, 33(3), 278–286.

Oweis, I.S. & Khera, R.P. (1998). *Geotechnology of Waste Management*. PWS, Boston.

Ozcep, F., Tezel, O. & Asci, M. (2009) Correlation between electrical resistivity and soil-water content: Istanbul and Golcuk. *International Journal of Physical Sciences*, 4(6), 362–365.

Pacey, J., Augenstein, D., Morck, R., Reinhart, D., & Yazdani, R. (1999). The bioreactor landfill-an innovation in solid waste management. *MSW Management*, 53–60.

Pánek, T., Margielewski, W., Táborík, P., Urban, J., Hradecký, J. & Szura, C. (2010) Gravitationally induced caves and other discontinuities detected by 2D electrical resistivity tomography: Case studies from the Polish Flysch Carpathians. *Geomorphology*, 123(1), 165–180.

Pohland, F.G. (1975) Sanitary landfill stabilization with leachate recycle and residential treatment. *EPA Grant No. R-801397*, U.S.E.P.A. National Environmental Research Center, Cincinnati.

Pozdnyakov, A.I., Pozdnyakova, L.A., & Karpachevskii, L.O. (2006) Relationship between water tension and electrical resistivity in soils. *Eurasian Soil Science*, 39(1), 78–83.

Qian, X., Koerner, R.M. & Gray, D.H. (2002) *Geotechnical Aspects of Landfill Design and Construction*. Prentice Hall, Upper Saddle River, NJ.

Reinhart, D.R. & Townsend, T.G. (1998) *Landfill Bioreactor Design and Operation*. Lewis, New York.

Reinhart, D.R. & Townsend, T.G. (2007) Update on Florida NRRL bioreactor landfill demonstration project. Presented at the *Illinois Recycling & Solid Waste Conference and Trade Show*.

Revil, A., Cathles, L.M., Losh, S. & Nunn, J.A. (1998) Electrical conductivity in shaly sands with geophysical applications. *Journal of Geophysical Research: Solid Earth*, 103(B10), 23925–23936.

Revil, A., Darot, M. & Pezard, P.A. (1996) From surface electrical properties to spontaneous potentials in porous media. *Surveys in Geophysics*, 17(3), 331–346.

Richardson, E.V., and Davis, S.R. (2001). Evaluating scour at bridges, 4th ed.: U.S. Department of Transportation, Federal Highway Administration, Hydraulic Engineering Circular 18, Publication FHWA NHI 01-001, 378 p.

Rinaldi, V.A. & Cuestas, G.A. (2002) Ohmic conductivity of a compacted silty clay. *Journal of Geotechnical and Geoenvironmental Engineering*, 128(10), 824–835.

Robain, H., Descloitres, M., Ritz, M. & Atangana, Q.Y. (1996) A multiscale electrical survey of a lateritic soil system in the rain forest of Cameroon. *Journal of Applied Geophysics*, 34(4), 237–253.

Robinson, W. (1993) Testing soil for corrosiveness. *Materials Performance*, 32(4), 56–58.

Saarenketo, T. (1998) Electrical properties of water in clay and silty soils. *Journal of Applied Geophysics*, 40(1–3), 73–88.

Sadek, M.S. (1993) *A Comparative Study of the Electrical and Hydraulic Conductivities of Compacted Clay*. Ph.D. Dissertation, University of California, Berkeley.

Saleh, A.A., Wright, S.G., 1997. *Shear strength correlations and remedial measure guidelines for long-term stability of slopes constructed of highly plastic clay soils*. Research Report 1435-2F. Center for Transportation Research, Bureau of Engineering Research. University of Texas at Austin, Austin, TX.

Samouelian, A., Cousin, I., Tabbagh, A., Bruand, A. & Richard, G. (2005) Electrical resistivity survey in soil science: A review. *Soil Tillage Research*, 83(2), 173–193.

Santamarina, J.C., Klein, K.A. & Fam, M.A. (2001) *Soil and Waves: Particulate material behavior, characterization and monitoring*. John Wiley and Sons, New York.

SAS Institute Inc. (2009) SAS® 9.2 macro language: Reference. *SAS Institute Inc.*, Cary, NC.

Sauer Jr., M.C., Southwick, P.E., Spiegler, K.S. & Wyllie, M.R.J. (1955) Electrical conductance of porous plugs: Ion exchange resin-solution systems. *Industrial and Engineering Chemistry*, 47, 2187–2193.

Schwartz, B.F., Schreiber, M.E. & Yan, T. (2008) Quantifying field-scale soil moisture using electrical resistivity imaging. *Journal of Hydrology*, 362(3–4), 234–246.

Shah, P. & Singh, D. (2005) Generalized Archie's law for estimation of soil electrical conductivity. *Journal of ASTM International*, 2(5), 1–19.

Shihada, H. (2011) *A Non-Invasive Assessment of Moisture Content of Municipal Solid Waste in a Landfill Using Resistivity Imaging*. Ph.D. Dissertation, University of Texas at Arlington.

Shihada, H., M. Hossain, S., Kemler, V. & Dugger, D. (2013) Estimating moisture content of landfilled municipal solid waste without drilling: Innovative approach. *Journal of Hazardous, Toxic Radioactive Waste*, 17, 317–330.

Simunek, J., Sejna, M., Saito, H., Sakai, M., & van Genuchten, M.T. (2002) *The Hydrus 1D Software Package for Simulating the One-Dimensional Movement of Water, Heat, and Multiple Solutes in Variably Saturated Media*. Department of Environmental Sciences, University of California Riverside, California.

Sposito, G. (2008) *The Chemistry of Soils*. Oxford University Press, USA.

Tabbagh, A. & Cosenza, P. (2007) Effect of microstructure on the electrical conductivity of clay-rich systems. *Physics and Chemistry of the Earth, Parts A/B/C*, 32(1), 154–160.

Taufiq, T. (2010) *Characteristics of Fresh Municipal Solid Waste*. Master's Thesis, University of Texas at Arlington.

Texas Commission of Environmental Quality. (2005) *Lake Gladewater Dam Inspection Report*. Inspection Report, Austin, TX.

U.S. Geological Survey. (2014). Texas geologic map data. Available from http://mrdata.usgs.gov/geol ogy/state/state.php?state=TXN July 25, 2014.

Van Nostrand, R.G. & Cook, K.L. (1966) Interpretation of resistivity data. *Geological Survey*, Professional paper; 499. U.S. Govt. Print. Off., Washington, 310.

Warith, M. (2002) Bioreactor landfills: Experimental and field results. *Waste Management*, 7–17.

Waxman, M.H. & Smits, L.J.M. (June, 1968) Electrical conductivity in oil bearing shaley sands. *Society of Petroleum Engineering Journal*, 107–122.

Yang, J.S. (2002) *Three Dimensional Complex Resistivity Analysis for Clay Characterization in Hydrogeologic Study*. Ph.D. thesis, University of California, Berkeley.

Yesiller, N., Hanson, J. & Liu, W. (2005) Heat generation in municipal solid waste landfills. *Journal of Geotechnical and Geoenvironmental Engineering*, 131(11), 1330–1344.

Yoon, G.L. & Park, J.B. (2001) Sensitivity of leachate and fine contents on electrical resistivity variation of sandy soils. *Jounal of Hazardous Materials*, 84(2–3), 147–161.

Yukselen, Y. & Kaya, A. (2006) Prediction of cation exchange capacity from soil index properties. *Clay Minerals*, 41(4), 827–837.

Zha, F., Liu, S. & Du, Y. (2007) Evaluation of change in structure of expansive soils upon swelling using electrical resistivity measurements. In: *Advances in Measurement and Modeling of Soil Behavior*. ASCE. pp.1–10.

Index

T - #0184 - 111024 - C244 - 246/174/12 - PB - 9780367571245 - Gloss Lamination